Economics of Production and Marketing of Citrus

Anil Bhat
Jyoti Kachroo
Division of Agricultural Economics and Statistics
Sher-e-Kashmir University of Agricultural Sciences and Technology of Jammu
Chatha, Jammu – 180009

2012
DAYA PUBLISHING HOUSE®
New Delhi – 110 002

ISBN 978-81-7035-800-8

Published by : **Daya Publishing House®**
4760-61/23, Ansari Road, Darya Ganj,
New Delhi - 110 002
Phone: 23245578, 23244987
Fax: (011) 23260116
e-mail : dayabooks@vsnl.com
website : www.dayabooks.com

Laser Typesetting : **Classic Computer Services**
Delhi - 110 035

Printed at : **Chawla Offset Printers**
Delhi - 110 052

PRINTED IN INDIA

Preface

Citrus is one of the major commercial fruit crops grown in India. India is the fifth largest producer of citrus fruits in the world. In terms of area under cultivation in India, citrus is the third largest fruit industry after banana and mango. Over the last 30 years, the area and production under citrus cultivation has increased at the rate of 11 per cent and 9 per cent, respectively. In India, it occupies an area of about 0.81 million ha with production of 7.50 million tonnes and productivity of 9.26 tonnes/ha. Among the citrus group, orange, kinnow and lemon is the most important fruit crop in India. Orange occupies an area of about 0.21 million ha with production of 1.44 million tonnes and productivity of 6.7 tonnes/ha whereas lemon occupies an area of 0.29 million ha with production of 2.43 million tonnes and productivity of 8.5 tonnes/ha. As far as export of citrus fruit is concerned, orange is main fruit crop which is exported to various countries such as Bangladesh, Nepal, USA, Canada etc. Export of orange outside the country during 2007-08 was 29.26 million tonnes. The foreign exchange earnings from export of orange were ₹27.27 crore for 2007-08.

In the State of Jammu and Kashmir, the area under citrus production is 11806 hectares with a production of 19208 metric tonnes. In the Jammu region of the J and K State, it occupies an area of 11762 hectares which is 99.62 per cent of total area of citrus in J and K with a production of 19202 metric tonnes. The Jammu region is predominantly rain-fed as only about 25 per cent area is irrigated. The Jammu region is suitable for horticultural crops especially for growing citrus fruits. With the rapid increase in the area under the crop in the Jammu region of J and K state, several problems of production and marketing have emerged which needed careful investigation. Efficient marketing plays an important role in the development of any enterprise. Hence, it was found necessary to investigate the prevalent marketing systems and channels, the marketing costs, margins and price spread in different channels as well as in different markets and other general problems faced by the

citrus growers in selling their produce which thereby can help the policy planners for its improvement in production and marketing strategy. In this book, an attempt has been made to investigate production and marketing aspects of the citrus cultivation.

First chapter provides an overview of citrus production. Second chapter deals with laws of production and past research work related to production aspects. Theory of marketing and brief research work carried so far in India and abroad concerning to marketing aspects are detailed in chapter third. Chapter fourth deals with a case study of citrus production and marketing in the Jammu region. Fifth and last chapter presents the future strategies for the production and marketing of citrus.

We acknowledge our heart-felt gratitude to Dr. Rajinder Peshin, Associate Professor Division of Agricultural Extension Education, Sher-e-Kashmir university of Agricultural Sciences and Technology of Jammu, Main Campus, Chatha, for his guidance, valuable criticism, useful comments and editing of the manuscript. We would remiss if we do not mention the name of Prof. Dileep Kachroo (Head, Agronomy, Chief Scientist, Farming System Research Centre) for his inspiring support, constructive suggestions and timely help in completing the book. His valuable suggestions were complementary to our research. We are thankful to Mr. Pawan Kumar Sharma (SMS, KVK, Poonch) for his help in one form or the other.

I (First Author) express my deep feelings to my dear sisters Sunita Goja, Rita Raina and Anita Raina and Brothers in law Sh. Bimal Goja, Sh. Shiban Krishan Raina and Sh. Vinod Raina for their moral support. I also express my regards and gratitude to my mother Smt. Kaushalya Bhat, who always showered her love and affection, and I am also grateful to my wife, Mrs. Sanju Bakshi for her continuous encouragement. My source of aspiration is my daughter, Baby Aaishani Bhat. I feel highly privileged, when I award them with the epithet of peerless asset, who revitalized my dream to live and theme to struggle with an eternal touch of dedication.

We convey our special thanks for Mr. Anil Mittal of Daya Publishing House, New Delhi who have taken sincere efforts to compile this book and bring in the nice form.

Anil Bhat

Jyoti Kachroo

Contents

List of Figures

List of Tables

Chapter 1

Citrus Production: An Overview

Agriculture continues to be core sector of the Indian economy, on which more than 60 per cent of our population is dependent for their livelihood. Agriculture, the most important industry of India, contributes about 14.6 per cent of the national income and employs directly or indirectly 52.1 per cent of its population (Anonymous, 2010) thereby indicating its predominance. But unfortunately Indian farmers in general are small and marginal land holders with poor resources and realizing income mainly from cereal based production system, which is insufficient to improve their living standards. According to the Indian Council of Medical Research (ICMR), a balanced diet should have nearly 300 grams of cereals, 60 grams of pulses, 280 grams of vegetables including tubers and 90 grams of fruits per day (Mruthyunjaya, 2001). A clear-cut increase in their consumption in recent years is visible particularly in urban areas on account of higher income. Therefore, lot of diversification has taken place in which fruits have become the significant part of one's diet. In such a typical scenario, need of the hour is to adopt such production system, which is capable of saving the interest of small farmers and increasing their economic condition. Under present agricultural scenario, fruits in the food consumption play the key role in enhancing trade and business worldwide. The country's population has almost tripled in the last five decades and its food grain production has more than quadrupled, significantly enhancing the per capita food grain availability. On the other hand, the share of agriculture in GDP has declined substantially from 55 per cent in early 1950s to about 42 per cent in the 1980s, and further to 14.6 per cent in 2009-10 (Anonymous, 2010). Now crop diversification has become the slogan of overall and sustainable agricultural development. So, the government has identified horticultural crops as a means of diversification of agricultural crops, more profitable on the basis of efficient land use, optimum utilization of natural resources and creating employment opportunities for rural masses especially women folk.

Citrus fruits are produced all over the world. According to UNCTAD, in 2004 there were 140 citrus producing countries. Around 70 per cent of the world's total citrus production is grown in the Northern Hemisphere, in particular countries around the Mediterranean and the United States, although Brazil is also one of the largest citrus producers. In India, in terms of area under cultivation, citrus is the third largest fruit industry after Banana and Mango. Over the last 30 years, the area and production under citrus cultivation has increased at the rate of 11 per cent and 9 per cent, respectively, which shows that the expansion of citrus industry was quite sustainable. Citrus is grown in more than 26 states in the country. It occupies an area of about 0.81 million ha with production of 7.50 million tones and yield of 9.26 tones/ha in India and ranks fifth in its production in the world (Anonymous 2010).

1.1. Origin and Uses of Citrus Fruit

While the origin of citrus fruits cannot be precisely identified, researchers believe they began to appear in Southeast Asia around 4000BC bordered by Northeastern India, Myanmar (Burma) and the Yunnan province of China. From there, they slowly spread to Northern Africa, mainly through migration and trade. During the period of the Roman Empire demand by higher-ranking members of society, along with increased trade, allowed the fruits to spread to southern Europe. Citrus fruits spread throughout Europe during the Middle Ages, and were then brought to the Americas by Spanish explorers. Worldwide trade in citrus fruits didn't appear until the 19th century and trade in orange juice developed as late as 1940 (Webber, 1967). The generic name originated in Latin, where it specifically referred to the plant now known as Citron (*C. medica*). It was derived from the ancient Greek word for cedar, (*kedros*). Some believe this was because Hellenistic Jews used the fruits of *C. medica* during Sukkot (Feast of the Tabernacles) in place of a cedar cone (Kimball, 1999), while others state it was due to similarities in the smell of citrus leaves and fruit with that of cedar (Spiegel *et al.*, 1996). Collectively, *Citrus* fruits and plants are also known by the Romance loanword agrumes (literally "sour fruits").

Citrus is one of the major commercial fruit crop widely consumed both as fresh fruit and juice. The term citrus fruit includes different types of fruits and products. Although oranges are the major fruit in the citrus fruits group, accounting for about 70 per cent of citrus output, the group also includes small citrus fruits (such as tangerines, mandarines, clementines and satsumas), lemons and limes and grapefruits. The leading processed form in the group is orange juice. Its global demand is attributed to its high vitamin C content and its antioxidant potential (Gorinstein *et al.*, 2001). It is rich in folic acid, a good source of fiber, fat free, sodium free and cholesterol free with additional quality of containing potassium, calcium, folate, thiamin, niacin, vitamin B_6, phosphorus, magnesium and copper. It may help to reduce the risk of heart diseases and some types of cancer and is also helpful to reduce the risk of pregnant women to have children with birth diseases (Economos *et al.*, 1999). Even essential and volatile oils are obtained from the citrus fruits peel sacks. So much so, it is used by the food industry to give flavor to drinks, foods and also acts as a component for the pharmaceutical industry for the preparation of medicines, soaps, perfumes and other cosmetics, as well as for home cleaning products

(Braddock, 1999). Citrus is also having its value added products such as Lime Chutney, Pudding, Juice (orange/lemon/lime), pickles, jam, jelly, marmalade, wine, vinegar, and confectionaries. Approximately one third of total citrus production is utilized for processing. This proportion is higher in the case of oranges as more than 40 per cent of globally produced oranges are utilized for processing. In addition, orange utilization for processing accounts for more than 80 per cent of total citrus utilization for processing. The proportion of grapefruit utilization for processing is similar to that of orange. In contrast, nearly all small citrus fruits of the tangerine type are intended for consumption in the fresh market. Lemons and limes are somehow different since they are normally consumed in association with other food products. They are grown mainly for the fresh market and their juice is used primarily as a flavoring in beverages (UNCTAD, 2008).

Citrus fruits flourish well on light soils with a good drainage. Deep soils with pH range of 5.5 to 7.5 are considered good. However, they can grow in pH range of 4 to 9. Light loam on heavier but well drained sub-soils appears to be ideal for citrus. Generally, citrus trees start bearing fruits 3 - 5 years from planting (although economic yields start from the fifth year and the trees may take 8 to 10 years to achieve full productivity) and can be harvested 5 - 6 months from flowering depending on the variety and the environment. Only a small percentage of flowers produce fruits. Citrus trees require a rich, well-drained soil. Citrus growing needs periodical fertilization and irrigation of the soil, as well as pruning of the tree.

1.2. World Citrus Production

Citrus fruits rank first in international fruit trade in terms of value. As a result of trade liberalization and technological advances in fruit transport and storage, the citrus fruit industry is becoming more global in scope. During the last decades, production and trade in citrus fruits have increased steadily, although the intensity of growth has been different according to the type of fruit (it has been stronger in small fruits and orange juice, particularly in not-from-concentrated orange juice during the nineties, while there is a certain stagnation of other citrus fruits consumption in developed countries). According to the United Nations Conference on Trade and Development (UNCTAD), the rise in citrus production is mainly due to the increase in cultivation areas, improvements in transportation and packaging, rising incomes and consumer preference for healthy foods. There is an increasing presence of counter-season citrus fruits going from the Southern Hemisphere to the Northern Hemisphere, contributing to year-round availability of fruit in the consuming areas in the North. Large Asian consumer markets are also opening new prospects for future trade expansion in citrus fruits (Economos *et al.*, 1999).

It is the fruit of sub-tropical and tropical regions of the world between 40° North and South latitude in over 137 countries on six continents (Ismail and Zhang, 2004). However, the sub-tropical climate is the best suited for citrus growth and development. Temperature below –4°C is harmful for the young plants. Soil temperature around 25°C seems to be optimum for root growth. Dry and arid conditions coupled with well defined summer having low rainfall (ranging from 75 cm to 250 cm) are most favourable for the growth of the crop. Citrus is considered as one of the important

fruit crop in international trade next to grapes having excellent quality and shelf life attributes and generate almost about 105 billion dollar per year in the world fruit market. This fruit with its origin in India (North East India is the native place of many citrus places) and China covers an area of about 7.37 million ha with production of 108 million tones and yield 14.72 tones/ha in the world. The major citrus producing countries are Spain, USA, China, India, Iran, Mexico, Italy, Nigeria, South Africa, Japan, Brazil, Turkey and Cuba. The table 1.1 explains the world top citrus producers.

Table 1.1: Top Ten Total Citrus Producers (tonnes)

Country	*Grapefruit*	*Lemons and Limes*	*Oranges*	*Tangerines*	*Other*	*Total*
Brazil	72,000	1,060,000	18,279,309	1,271,000	-	20,682,309
China	547,000	745,100	2,865,000	14,152,000	1,308,000	19,617,100
United States	1,580,000	722,000	7,357,000	328,000	30,000	10,017,000
Mexico	390,000	1,880,000	4,160,000	355,000	66,000	6,851,000
India	178,000	2,060,000	3,900,000	-	148,000	6,286,000
Spain	35,000	880,000	2,691,400	2,080,700	16,500	5,703,600
Iran	54,000	615,000	2,300,000	702,000	68,000	3,739,000
Italy	7,000	546,584	2,293,466	702,732	30,000	3,579,782
Nigeria	-	-	-	-	3,325,000	3,325,000
Turkey	181,923	706,652	1,472,454	738,786	2,599	3,102,414
World	**5,061,023**	**13,032,388**	**63,906,064**	**26,513,986**	**7,137,084**	**115,650,545**

Source: Food and Agricultural Organization of United Nations: Economic and Social Department: The Statistical Division

However, the biggest citrus exporters in descending order are:

Spain, South Africa, United States, Turkey, Argentina, China, Mexico and Morocco.

The biggest citrus importers by far are the 27 EU countries whose main EU external imports come from South Africa, Argentina and Turkey. After the EU the biggest importers are Russia, United States (orange juice, lemons, limes and mandarins), Canada, Japan, Ukraine, Hong Kong, Malaysia, Switzerland and Indonesia. (USDA Foreign Agricultural Service: Citrus World Markets and Trade, 2008).

1.3. Fruit Production in India

Indian soil bestowed with a wide range of climates and physio-geographical conditions is most suitable for growing various kinds of fruits, flowers, vegetables, nuts, spices and plantation crops. Horticulture crops cover only 18.98 million hectares *i.e.*, 13.27 per cent of total cultivated area (7.2 million hectare under vegetables, 0.15 million hectare under loose flowers, 3.2 million hectare under plantation crops, 2.4 million hectare under spices and 5.51 million hectare under fruits) but contribute 28 per cent of the gross domestic product in agriculture (Mittal, 2007). Not only this, it is

the world's second largest producer of fruits (71.51 million tones) with its projected value of touching 98 million tones by the year 2020-2021 (Banerjee, 2009) and vegetables (133.74 million tonnes) and contributes 11 per cent and 14 per cent, respectively, in the world (Anonymous, 2011). Productivity of horticultural crops has increased from 1.2 tonnes per hectare in 1953-54 to around 4.95 tonnes per hectare in 2007-08. About 65-70 per cent of fruits and vegetables produced in India are consumed domestically, two per cent of them are being processed, and only one per cent is being exported while post harvest losses account to 20-30 per cent of the stored fruits (Wani, 2008). India's export of fresh fruits and vegetables has increased from US$ 1713.00 million in 2006-07 to US$ 2093.00 million in 2008-09 *i.e.*, an increase of 22.18 per cent (Anonymous, 2011a).

Among the fruit crops, citrus is of wide importance in India. In India, citrus fruits are primarily grown in Maharashtra, Andhra Pradesh, Punjab, Karnataka, Uttaranchal, Bihar, Orissa, Assam and Gujarat. Of the various types of citrus fruits grown such as Lemons, Oranges, Mandarins, Grape fruits, Tangelos, Pummelo, Kumquats, Limes, Citrons, Lavender gem, Oroblanco, Shaddock, Tangerine, Ugli fruit; orange (mandarin or santra), sweet orange (mosambi, malta or satgudi) and lime/lemon are of commercial importance in India. Mandarin orange with certain specifications is exported from India to other countries, which are given in Tables 1.2 and 1.3, respectively.

Table 1.2: Export Specifications for Mandarin Orange

Variety	*Nagpur Mandarin*
Fruit Colour	Light orange
Fruit Weight	150-175 Gms
Grades	65-70 mm and 40-45 mm
Packing 65 mm grade	7 Kg
40 mm grade	2.5 Kg
Storage	5-7° C
Transport	By Sea

Source: Maharashtra State Agriculture Marketing Board Website (2009).

Oranges keep well for a long time under ambient conditions and hence can be transported to distant places for marketing. Citrus fruits are sold throughout the country. Several fruit processing units also purchase citrus fruits in bulk. Indian sweet oranges are exported to France, UK, Belgium, Indonesia, Netherland, SriLanka, Bangladesh and many other countries. Limes are exported to UAE, Saudi Arabia, Bangladesh and few other countries (http://fruitsexotic.blogspot.com).

1.4. Citrus Production in Jammu and Kashmir

The state of Jammu and Kashmir with its favourable climatic conditions for horticultural crops is suitable for growing many fruits of commercial importance. Agro-climatic suitability equips the state of Jammu and Kashmir with a unique comparative edge in the cultivation of a variety of horticultural crops. Apple, pear, citrus, mango, olive, apricot, peach, grapes, ber, walnut, almond etc are the fruits mostly grown in this state. The state accounts for sixty per cent production of the apple in the country and is also known as the walnut state of the country. The state has got the monopoly of trade in cherry (Kachroo, 2004).

Table 1.3: Export of Orange from India

Value in Rupees
Quantity in kg

Country Name	Qty. 2005-06	Value 2005-06	Qty. 2006-07	Value 2006-07	Qty. 2007-08	Value 2007-08
Bangladesh	34955079	335138808	27958552	272176931	26002070	244426144
Nepal	1319025	10436306	1491782	9033940	2615554	17494094
USA	1000	20031	0	0	348000	5032922
Canada	0	0	0	0	113000	2721861
Netherlands	0	0	0	0	40000	1206269
Oman	14140	752828	3984	117243	35482	439335
Spain	0	0	0	0	24000	344632
China	0	0	0	0	31283	334516
UAE	36419	620932	144250	2068159	25600	332900
Malaysia	0	0	0	0	4000	166353
Others	161900	1755634	511475	7990012	22311	184046
Total	**36487563**	**348724539**	**30110043**	**291386285**	**29261300**	**272683072**

Source: APEDA (2009).

The fruit industry is the second most important industry after tourism in Jammu and Kashmir state with 41.82 per cent area under horticultural crops out of the total net sown area of 7.34 lakh hectares and an income of ₹2800.00 crore has been generated from fruits production during 2009-10 (₹300 crore from dry fruits) (Anonymous, 2011b).

Almost 45 per cent of economic returns in agriculture sector is accounted for by horticulture showing its growing importance in the state economy. Its contribution to GSDP has been estimated to be 7-8 per cent. For the year 2008-09, an amount of ₹5796.49 lakh was earmarked on horticulture sector both under Plan and Non-plan budget. The projected amount being spent on this sector during 2009-10 works out to ₹7929.18 lakh both under plan and non-plan budget. The horticulture sector provides remunerative means for diversification of land use for improving productivity and returns. It increases employment opportunities and earns foreign exchange. It also provides nutritional security and raw material for growing agro processing industries. The fruit industry in J&K state involves about five lakh orchardists in its trade, generates employment with 400 man days per hectare per year and provides direct and indirect employment to over 30 lakh people (Anonymous, 2009). Due to potential demand for temperate fresh and dry fruits, the state recorded the fruit export of 11.16 lakh metric tonnes during 2008-09 of which 11.01 lakh tonnes (98.65 per cent) were fresh fruits. During 2008-09, the total fruit production was 16.91 lakh metric tonnes comprising of 15.30 lakh metric tonnes of fresh fruit and 1.61 lakh metric tonnes of dry fruit (Anonymous, 2009). The major fruits crops grown in the state are apple, citrus, mango, walnut, pear, grapes olive, ber, apricot, plum, peach etc. Citrus fruits are important fruit crops in rainfed and irrigated ecosystems of Jammu region.

Among the various fruit crops, growing of citrus has vast potential in Jammu region of Jammu and Kashmir state as it comprises highest area under its cultivation (11762 hectares) which is 99.62 per cent of total area of citrus in J and K, whereas its production has been realized to (19202 metric tonnes) which is 99.96 per cent of the total production of J and K (Anonymous, 2009). In Jammu region, the districts mainly Rajouri, Kathua, Jammu, Udhampur, Samba and Reasi are the prominent areas where it is grown.

1.5. Conclusion

Citrus occupies a place of importance in the horticultural wealth and economy of the world as well as India. Citrus fruits rank first in international fruit trade in terms of value. India is blessed with diverse environment conducive to the production of number of fruits. In India, in terms of area under cultivation, citrus is the third largest fruit industry after Banana and Mango. Over the last 30 years, the area and production under citrus cultivation has increased which shows that the expansion of citrus industry is quite sustainable. It occupies an area of about 0.81 million ha with production of 7.50 million tones and yield of 9.26 tones/ha in India and ranks fifth in its production in the world. Among all the states, Jammu and Kashmir is one of the major fruit producing states of India. The traditional hill economy of Jammu and Kashmir is slowly but steadily moving towards commercialized agriculture production. The state has made tremendous progress in fruit production over the period of last three decades. The area under citrus fruit was about 11762 hectares and its production was 19202 metric tonnes in Jammu region of Jammu and Kashmir state during the year 2008-09.

Chapter 2
Economic Theories of Production

Production refers to the economic process of converting of inputs into outputs. Production uses resources to create a good or service that is suitable for use, gift-giving in a gift economy, or exchange in a market economy. This can include manufacturing, storing, shipping, and packaging. Also level of output of a particular commodity depends upon the quantity of input used for its production. In other words production means transforming inputs (land, labour, capital) into an output.

2.1. Theoretical Orientation

2.1.1. Laws of Production

Theory of production is based on the following laws of production such as:

"Law of Diminishing Returns/Law of Increasing Cost, Law of Increasing Returns/Law of Diminishing Cost and Law of Constant Returns/Law of Constant Cost"

2.1.1.1. Law of Diminishing Returns/Law of Increasing Cost

The law of diminishing returns (also called the Law of Increasing Costs) is an important law of micro economics. The law of diminishing returns states that:

> *"If increasing amounts of a variable factor are applied to a fixed quantity of other factors per unit of time, the increments in total output will first increase but beyond some point, it begins to decline".*

Richard A. Bilas describes the law of diminishing returns in the following words:

> *"If the input of one resource to other resources are held constant, total product (output) will increase but beyond some point, the resulting output increases will become smaller and smaller".*

(a) Operation of Law of Diminishing Returns

The classical economists were of the opinion that the law of diminishing returns applies only to agriculture and to some extractive industries, such as mining, fisheries urban land, etc. The law was first stated by a Scottish farmer as such. It is the practical experience of every farmer that if he wishes to raise a large quantity of food or other raw material requirements of the world from a particular piece of land, he cannot do so. He knows it fully that the producing capacity of the soil is limited and is subject to exhaustation.

As he applies more and more units of labour to a given piece of land, the total produce no doubt increases but it increases at a diminishing rate. For example, if the number of labour is doubled, the total yield of his land will not be doubled. It will be less than double. If it becomes possible to increase the yield in the very same ratio in which the units of labour are increased, then the raw material requirements of the whole world can be met by intensive cultivation in a single flower-pot. As this is not possible, so a rational farmer increases the application of the units of labour on a piece of land up to a point which is most profitable to him. This in brief, is the law of diminishing returns. Marshall has stated this law as such:

> *"An increase in capital and labour applied to the cultivation of land causes in general a less than proportionate increase in the amount of the produce raised, unless it happens to coincide with the improvement in the act of agriculture."*

This law can be made clearer if we explain it with the help of a table and a curve.

Table 2.1: Total and Marginal Product in Response to Input (Labour)

Fixed Input	Variable Input	Total Product TP (in tons)	Marginal Product MP (in tons)
12 Acres	1 Labour	50	50
12 Acres	2 Labour	120	70
12 Acers	3 Labour	180	60
12 Acres	4 Labour	200	20
12 Acers	5 Labour	200	0
12 Acres	6 Labour	195	-5

In the Table 2.1, a firm first cultivates 12 acres of land (fixed input) by applying one unit of labour and produces 50 tonnes of wheat. When it applies 2 units of labour, the total produce increases to 120 tonnes of wheat, here, the total output increased to more than double by doubling the units of labour. It is because the piece of land is under-cultivated. Had he applied two units of labour in the very beginning, the marginal return would have diminished by the application of second unit of labour.

In the Table 2.1, the rate of return is at its maximum when two units of labor are applied. When a third unit of labour is employed, the marginal return comes down to 60 tonnes. With the application of 4th unit, the marginal return further, goes down to 20 tonnes and when 5th unit is applied it makes no addition to the total output. The

sixth unit decreased it. This tendency of marginal returns to diminish as successive units of a variable resource (labour) are added to a fixed resource (land) is called the law of diminishing returns. The Table 2.1 can be represented graphically in Figure 2.1.

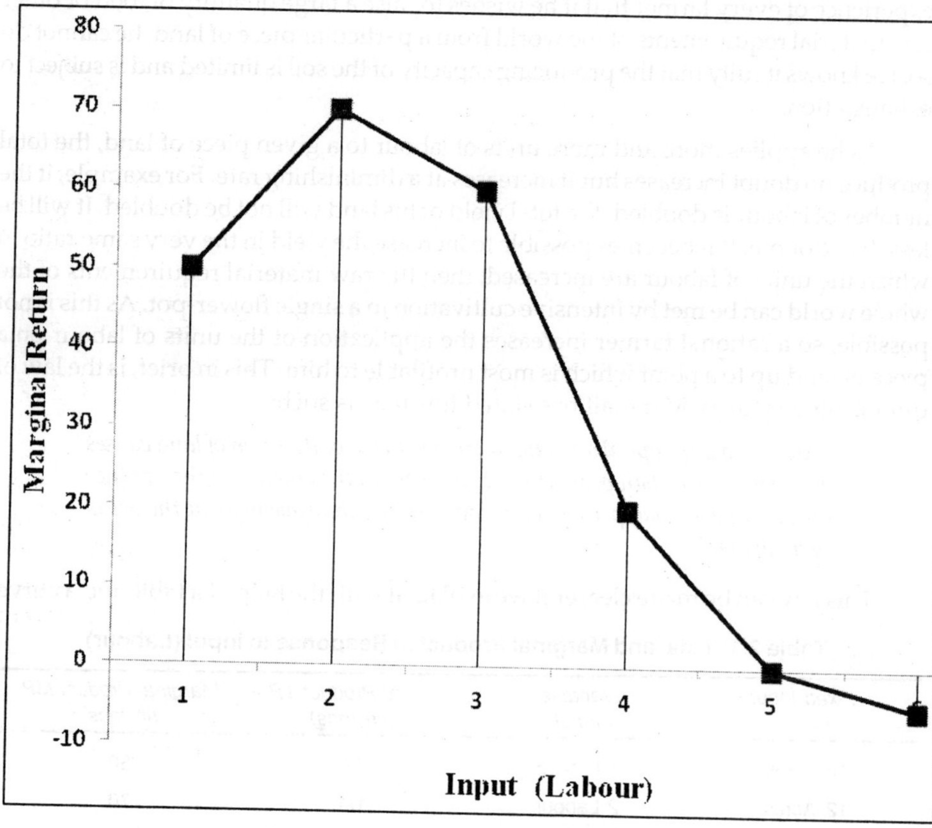

Figure 2.1: Law of Diminishing Returns

In Figure 2.1 along x-axis are measured doses of labour applied to a piece of land and along y-axis, the marginal return. In the beginning, the land was not adequately cultivated, so the additional product of the second unit increased more than of first. When 2 unit of labour was applied, the total yield was the highest and so was the marginal return. When the number of workers were increased from 2 to 3 and more, the MP began to decrease. As fifth unit of labour was applied, the marginal return fell down to zero and then it decreased by 5 tonnes.

(b) Assumptions

The Table 2.1 and the Figure 2.1 are based on the following assumptions:

1. The time is too short for a firm to change the quantity of fixed factors.
2. It is assumed that labour is the only variable factor. As output increases, there occurs no change in the factor prices.

3. All the units of the variable factor are equally efficient.
4. There are no changes in the techniques of production.

(c) Importance

The law of diminishing returns occupies an important place in economic theory. The British classical economists particularly Malthus, and Ricardo propounded various economic theories, on its basis. Malthus, the pessimist economist, has based his famous theory of Population on this law. The Ricardian theory of rent is also based on the law of diminishing return. The classical economists considered the law as the inexorable law of nature.

2.1.1.2. Law of Increasing Returns/Law of Diminishing Cost

The law of increasing returns is also called the law of diminishing costs. The law of increasing return states that:

> *"When more and more units of a variable factor is employed, while other factor remain fixed, there is an increase of production at a higher rate. The tendency of the marginal return to rise per unit of variable factors employed in fixed amounts of other factors by a firm is called the law of increasing return."*

An increase of variable factor, holding constant the quantity of other factors, leads generally to improved organization. The output increases at a rate higher than the rate of increase in the employment of variable factor.

Assumptions

The law rests upon the following assumptions:

1. There is a scope in the improvement of technique of production.
2. At least one factor of production is assumed to be indivisible.
3. Some factors are supposed to be divisible.

Example

The law of increasing returns can also be explained with the help of a table and a curve.

Table 2.2: Total Product and Marginal Product in Response to Input Application

Fixed Input	*Variable Inputs*	*Total Product*	*Marginal Product*
12 acres	1 labour	100	100
12 acres	2 labour	250	150
12 acres	3 labour	450	200
12 acres	4 labour	750	300
12 acres	5 labour	1200	450
12 acres	6 labour	1850	650

In the Table 2.2, it is clear that as the farmer goes on adding inputs, the total product goes on increasing with increasing return as is also clear from marginal product.

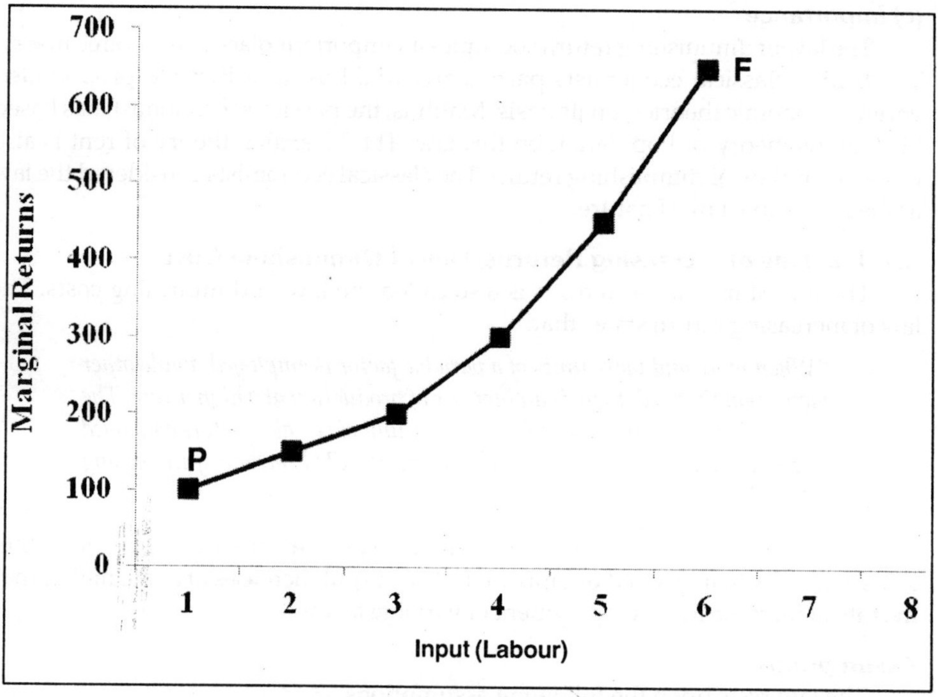

Figure 2.2: Law of Increasing Returns

In figure 2.2, along X-axis are measured the units of inputs applied and along Y-axis the marginal return is represented. PF is the curve representing the law of increasing returns.

2.1.1.3. Law of Constant Returns/Law of Constant Cost

The law of constant returns is also called as the law of constant cost. It is said to operate when with the addition of successive units of one factor to fixed amount of other factors, there arises a proportionate increase in total output. The yield of equal return on the successive doses of inputs may occur for a very short period in the process of production.

In the words of *Marshall*:

> *"If the actions of the law of increasing and diminishing returns are balanced, we have the law of constant return."*

Assumptions

In actual life, the law of constant returns can operate only if the following conditions are fulfilled:

1. There should not be any increase in the prices of inputs required.
2. The prices of various factors of production should remain the same.
3. The productive services should not be fixed and indivisible.

The law of constant returns can operate for a very short period when the marginal return moves towards the optimum point and begins to decline. If the marginal return, at the optimum level remains the same with the increased application of inputs for a short while, then we have the operation of law of constant returns. The law is represented now in the form of a table and a curve.

Table 2.3: Total Product and Marginal Product in Response to Input Application

Fixed Input	*Input Variable*	*Total Product*	*Marginal Product*
12 acres	1 labour	60	60
12 acres	2 labour	120	60
12 acres	3 labour	180	60
12 acres	4 labour	240	60
12 acres	5 labour	300	60

In the Table 2.3, the marginal product remains the same, *i.e.* 60 units.

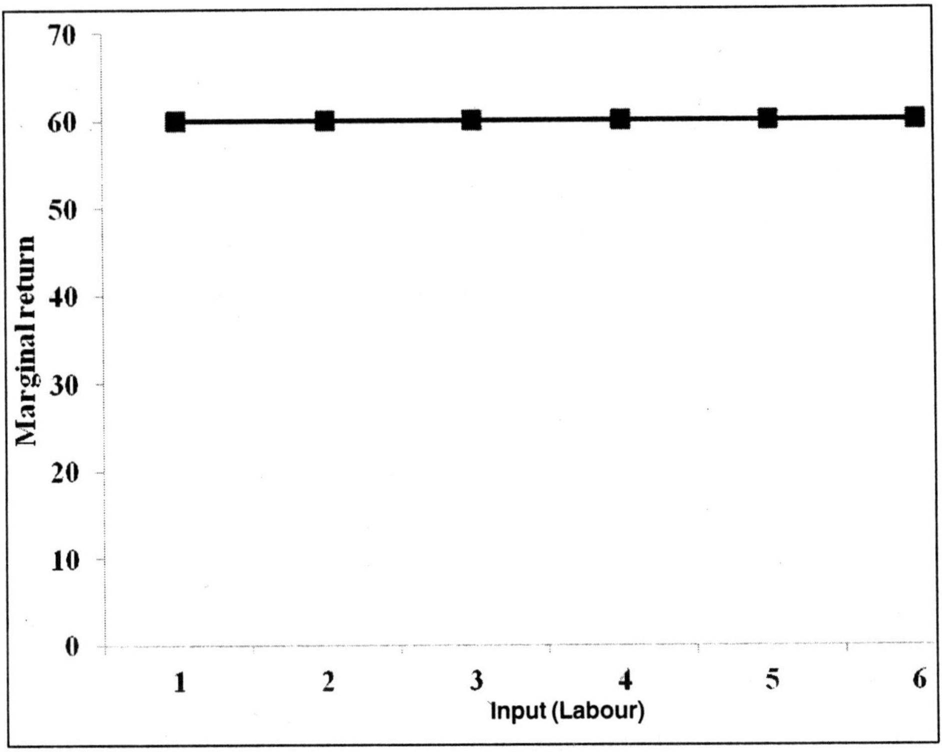

Figure 2.3: Law of Constant Returns

In Figure 2.3, along X-axis are measured the productive resources and along Y-axis is represented the marginal product. The straight line parallel to the X-axis represents the law of constant returns.

2.1.2. Production Function

The analysis of agricultural production is an integral part of agricultural development policy because of the vital position of agriculture in many developing countries (Yotopoulos *et al.*, 1979). A policy maker always requires information about the production responses of inputs and management practices before introducing any policy interventions to accomplish various short and long-term objectives. The purpose of this analysis is to identify the citrus fruits input-output relationship in the form of a mathematical function and to gain an understanding of the influence of the various inputs on output. Once such relationships are realized, then efficient use of inputs can be determined.

A production function is the core concept in the production theory of economics. It describes the physical relationship between input and output assuming the maximum output that can be produced from the input combination with given technology. Mathematically, it can be represented as:

$$Y = f(X, Z)$$

where,

f denotes the form of production technology. In the above equation Y, X and Z are vectors of non-negative output, variable and fixed inputs respectively. The production function rules out the possibility of negative output or input levels.

A production technology can be estimated in three equivalent ways *i.e.*, the production function, the profit function and the cost function. These three methods may be viewed as different roads leading to the same destination of maximizing profit. Duality theory establishes the correspondence between these three approaches. In economics, the Cobb-Douglas functional form of production functions is widely used to represent the relationship of an output to inputs. It was proposed by Knut Wicksell (1851–1926), and tested against statistical evidence by Charles Cobb and Paul Douglas in 1900–1928. Cobb-Douglas production function was the most popular specification both in theory and empirical analysis. This is mainly attributed to its simplicity, ease of estimation and interpretation. However, this function also assumes unit elasticity of substitution implying no complementarities between inputs, constant production elasticities, strictly linear expansion paths, strong separability and homogeneity (Heathfield and Wibe, 1987).

2.2. Review of Resource Use Efficiency Studies

Resource in agriculture means the proportion of the total stock that man can make available under technological and economic conditions for cultivation practices. The quantum of production under a normal crop season is directly related to the availability of resource input and their techniques of application. Productivity, the output flow per unit of resource input, depends on the level of input used, which in turn depends upon the investment pattern.

Wani *et al.* (1993) used Cobb-Douglas function to study the resource use efficiency of apple in Jammu and Kashmir. They divided the apple orchards into five groups on the basis of age of the orchards namely group A comprising of orchards of the age of

5-15 years, group B (16-23 years), group C (24-31 years), group D (32-39 years) and group E (≥ 40 years). The study revealed that most rationally used input was fertilizer which turned out to be positively significant in the three groups and also in overall group with regression coefficient as 0.257, while the most irrationally used input was plant protection which turned out to be positively significant (0.675) in only one group (32-39 years). Labour was found to be used rationally in one group. MVPs of fertilizer were positively significant in categories A, C and E and amounted to ₹64.40, ₹108.15 and ₹140.42, respectively, while MPVs of plant protection were positively significant in categories D and E amounting to ₹8563.35 and ₹6204.42, respectively. However, MVP of labour was positively significant in Category B where it amounted to ₹1164.25.

Chinappa and Ramanna (1997) used Cobb-Douglas production function to determine the level of resource use efficiency in Guava production in Allahabad. They included land, labour, manures, fertilizers and plant protection chemicals as the independent variables while yield was taken as the dependent variable. The functional analysis revealed that 42 per cent of the variation in the gross returns was explained by the independent variables under study. The regression coefficients of land and labour were significant while that of manures, fertilizers and plant protection measures were non significant.

Utomakili and Molua (1998) used Cobb-Douglas production function to determine the efficiency of resources used (fertilizer, labour, farm size) in banana production in Cameroon. Data were obtained from the annual production records of one of the major banana producing corporations in the country, the Cameroon Development Corporation, covering a period of 15 years. The results showed that there was inefficiency in input use, while labour input was over utilized, fertilizer use and farm size needed to be increased in order to improve resource use efficiency.

Koujalagi *et al.* (1999) examined the resource use efficiency in the cultivation of pomegranate in Bijapur district of Karnataka by using Cobb-Douglas type of production function. The function analysis revealed that 70 per cent of the total variation in gross returns was explained by the six variables *viz.* land, number of plants per acre, labour, plant protection chemicals, irrigation and manures and fertilizers. The regression coefficient of land, labour and manures and fertilizers indicated that the contribution of these inputs was significant, while as that of plants per acre, irrigation and plant protection chemicals was negative and non significant.

Naikwadi *et al.* (2004) carried out study on economics of production and marketing of fig in Pune district. Cobb-Douglas production function was used to determine level of resource use efficiency for fig crop for small, medium and large farmers of Pune district. Yield was considered as dependent variable, while human labour (man days), potassic fertilizers (kg), phosphatic fertilizers (kg), nitrogenous fertilizers (kg), expenditure on irrigation and manure (₹), expenditure on plant protection (₹) as independent variables. The analysis revealed that 64.29 per cent of variation in the output was explained by the selected seven explanatory variables. The regression coefficient of nitrogenous fertilizer was significant at 10 per cent level in I and II size group which indicated that there was a scope for increasing fig production by increasing the use of nitrogenous fertilizer.

Suresh and Keshava Reddy (2004) employed Cobb-Douglas type of production function to examine the productivity of resources for the banana crop of small, medium and large farmers of Thrissur district, Kerala. The independent variables included in the function were seedling, human labour, chemical fertilizer, organic manure, plant protection chemical and miscellaneous expenditure. They considered total returns from banana production as dependent variable. The productivity analysis indicated that the fertilizer, plant protection chemical and miscellaneous expenditure were significantly affecting the yield of banana. The elasticity coefficients for these variables were 0.16, -0.20 and 1.68 respectively. Resources like number of seedling per hectare of land, labour, organic manure and plant protection chemical were over applied. The allocative efficiency analysis indicated that there was scope for increasing the returns by increasing fertilizer and miscellaneous expenditure.

Zaman and Schumann (2005) studied the planted citrus groves at varying spacings to improve resource efficiency and to optimize fruit production for maximum economic return. Four commercial groves with different row spacings and tree ages were scanned with a Durand-Wayland ultrasonic system to measure and map tree volumes and to examine the effect of row spacings and tree ages on citrus yield. The ultrasonically measured volumes (UVs) were compared with manually measured tree volumes (MVs) of 30 trees in each grove to examine the performance of the ultrasonic system. The ultrasonic system measured tree volumes reliably in different groves with an average prediction accuracy (APA) >90 per cent , and correlation with manual measurement of $R^2 = 0.95–0.99$. Standard error of prediction and root mean square errors were relatively higher in widely spaced old groves than closely spaced young groves. The ultrasonically sensed tree volume map showed substantial variation in canopy volumes (0-240 m^3 $tree^{-1}$) within the grove. Therefore, the use of ultrasonic systems is a better option to quantify and map each tree volume rapidly (real-time) for planning site-specific management practices accurately in commercial groves and for estimating fruit yield.

Ahmad and Mustafa (2006) examined the forecasting kinnow production in Pakistan: An econometric analysis by using linear regression model. The functional analysis revealed that significance of the coefficients was a vital part of research findings. Estimated coefficients were found significant at one per cent level of significance. The value of R^2 was 0.83, which showed that 83.3 per cent variation in citrus production was explained by included explanatory variables.

Wagale *et al.* (2007) conducted a study to explore resource use efficiency in mango production and specifically to determine inputs utilization across different farm size, examine size productivity relationship in mango production. They found that on an average 67 per cent growers used manure, 70.83 per cent used fertilizers and 76.67 per cent used plant protection chemicals.

Hanumantharaya *et al.* (2009) conducted a study to compare resource use efficiency in sucker propagated banana and tissue culture banana. He employed Cobb-Douglas production function and estimated that in sucker banana cultivation, regression coefficient of plant nutrients (0.350) was significant at five per cent and that of plant protection chemicals (0.010) and bullock labour (0.048) were positive

but non-significant. The regression coefficient of human labour (0.078) was significant at one per cent. In tissue culture banana the regression coefficient of plantlets (0.091) was significant at five per cent level and coefficients of plant protection chemicals (0.020), human labour (0.120) and bullock labour (0.080) were significant at one per cent level. The R^2 value was 0.46 in case of tissue culture banana and 0.86 in case of sucker propagated banana which indicated that 46.0 and 86.0 per cent of variation, respectively, was explained by the independent variables considered in the model. The MVP of human labour, bullock labour and plant protection chemicals in tissue culture banana was positive with their values at 232.30, 60.93 and 2.19, respectively whereas in sucker propagated banana the MVP of human labour (1.10), plant protection chemicals (1.61) and plant nutrients (1.72) were found to be positive.

Iqbal (2009) conducted a study on investment appraisal of mango and ber fruit production in Jammu district of J and K state during the year 2008. He employed Cobb-Douglas production function and estimated that in mango cultivation, regression coefficient of human labour (0.453) was significant at five per cent and manures + fertilizers (0.674) at one per cent where as plant protection (0.009) was positive but non-significant. The regression coefficient of training/pruning (-0.003) was negative and non-significant. In ber the regression coefficient of human labour (1.119) was significant at one per cent level and that of training/pruning (0.004) was positive but non-significant. The R^2 value was 0.95 in case of mango and 0.90 in case of ber which indicated that 95.0 and 90.0 per cent of variation, respectively, was explained by the independent variables considered in the model. The MVP of human labour, manures + fertilizers and plant protection in mango was positive with their values at 2.622, 19.982 and 1913, respectively whereas in ber the MVP of human labour (18.594) and training/pruning (7.181) were found to be positive.

Khushk *et al.* (2009) conducted a study at Technology Transfer Institute (PARC), Tandojam, Pakistan during 2005-06 year and employed Cobb-Douglas production function to examine the productivity of resources for the Guava crop. The results showed that R^2 value was 0.68 which indicated that 68.0 per cent variation in guava production was explained by included explanatory variables. Number of inter-culturing (0.541) and soil type (0.223) were significant at 1 per cent level, whereas, coefficients of number of guava trees (0.172), farmyard manure (FYM) (0.162), fertilizer (0.192) and human labour (0.145) were significant at 5 per cent level of significance. Area of farmland devoted to guava production was not significant. Moreover, major production inputs such as pesticide sprays, fertilizer, FYM and labour for orchard management were underutilized, affecting guava production.

Landge *et al.* (2010) conducted the study during the year 2008-09 in Nanded district of Maharashtra to know the resource productivity, resource use efficiency and optimum resource use in banana production. He employed Cobb-Douglas production function and revealed that marginal productivity with respect to area, machine labour and bullock labour was 40.410, 2.615 and 1.867 quintals respectively. It inferred that if area was increased by one hectare, machine labour increased by one hour and bullock labour increased by one pair, it would lead to increase banana production by 40.410, 2.615 and 1.867 quintals, respectively.

It can be concluded that the regression coefficients of human labour, manures + fertilizers and land were significant which could be analysed by using Cobb-Douglas production function. The functional analysis revealed that significance of the coefficients was a vital part of research findings. The value of R^2 showed that variation in fruit production was explained by included explanatory variables.

2.3. Review of Costs and Returns Studies

In the farm business, knowledge of cost concept enables the cultivators to adjust and coordinate production resources for their profitable use.

Subrahmanyam (1986) in his study found that lime cultivation was profitable than the sweet orange. The data was collected from 60 cultivators of West Godawari district of Andhra Pradesh. The maintenance cost up to bearing was worked out to be ₹260.00 per hectare for sweet orange as compared to ₹476.00 in lime. The net returns realized were ₹6617.25 for lime and ₹5102.32 for sweet orange.

Thakur *et al.* (1986) studied the cultivation costs and net returns of Kinnow (hybrid mandarin) and concluded that it has become an important cash crop in Himachal Pradesh. Data was collected from 100 growers. The operational costs including replacement costs, fertilizers, chemicals, pruning, labour and irrigation were worked out to be ₹2172.14 per ha in the first year and increased annually upto ₹3416.22 in the tenth year. Total fixed cost, including amortized establishment costs for 25 years, was worked out to be approximately ₹872.00. The gross returns in the first harvest was found to be ₹5112.80 whereas net returns ranged from ₹1345.74 to ₹26494.48 between 4th year and 10 year.

Reddy (1987) worked out the economic feasibility of mango, acid lime and sweet lime and compared the economics of these crops with those of the arable crops grown in the area. The benefit cost ratio had been found higher for horticultural crops than for the competing arable crops. He had advocated the horticultural development programme in the state and should be prepared on the basis of land use capability, nature of commodity in relation to proximity of market, sound programme of financial assistance and existing market structure.

Sudha *et al.* (1988) conducted the study on economics of sweet orange cultivation in Cuddapah district of Andhra Pradesh. The year wise cost of cultivation was collected from 21 farmers and average was taken. The expenses on manures and fertilizers increased with age until the crop had come to the bearing stage. Plant protection (₹1050.00 per hectare) and watch and ward (₹1500.00 per hectare) were the highest cost on bearing orchards. The average annual cost of maintaining a bearing orchard was ₹7446.00 per hectare. The gross return in the first harvest ranged between ₹2053.00 to ₹3786.00 per hectare. The gross return showed a steady increase from ₹9206.00 per hectare in the sixth year to ₹13043.00 in the fifteenth year of the crop.

Hugar *et al.* (1991) found that the total establishment cost of guava orchard in Karnataka was ₹6448 per hectare, out of which the opportunity cost of land (46.45 per cent) formed the major component followed by the expenses on layers and grafts (12.54 per cent) and irrigation (8.67 per cent), hence the farmer owning the land had

only to incur 53.55 per cent of establishment costs. The maintenance cost from third year onwards varied from ₹2855 to ₹7034 per hectare while as gross and net returns per hectare varied from ₹2000 to ₹45400 and ₹856 to ₹38439 respectively. The maintenance cost, gross returns and net returns from 16th year onwards were ₹6578, ₹39586 and ₹33008 per hectare, respectively.

Satihal (1993) studied the costs and returns of ber in Bijapur district of Karnataka. The results revealed that investment for establishing one hectare of ber orchard was ₹97415.00 in medium and ₹99875.00 in large and small orchards respectively. The share of material cost in total investment was ₹48218.00 in small and ₹47232.00 in large orchards. Recurring and maintenance cost during gestation period was ₹50187.00 and ₹49659.00 in large and small orchards, respectively. Per hectare cost of cultivation during bearing period was relatively more in large orchards (₹77404.00) as compared to small orchards (₹66991.00).

Gangwar and Singh (1998) studied the economic evaluation of Nagpur mandarin cultivation in Vidharbha region of Maharashtra during 1996-97 and concluded that returns from Nagpur mandarin orchards started from sixth year and continued beyond 26 years. They found that the total establishment costs of orchard was ₹35452.00 per hectare and the maintenance costs from the sixth year onwards varied from ₹12667.00 to ₹22403.00 per hectare. Average gross returns during the bearing period amounted to ₹30864.00 per hectare.

Choubey and Atteri (2000) conducted the study during the year 1996-97 and found that the share of human labour in total establishment of litchi orchard, upto bearing stage was highest (52 per cent). The plant material cost constituted 13.46 per cent of the total investment and the share of manures + fertilizers together accounted 9.7 per cent. The share of plant protection material was 2.00 per cent. The share of investment on capital and rental value of land accounted for about 20 per cent of the total investment. He concluded that the total establishment cost upto 5 years in litchi in Bihar was ₹59326 per hectare with annual maintenance cost of ₹12184.

Radha *et al.* (2000) collected the data on costs and returns of citrus from the time of establishment till bearing from a sample of 36 orchards from six villages of Northern Telangana Zone of Andhra Pradesh and found that total costs were phenomenally higher in the first year in the pre bearing period and the average bearing cost was still higher accounting for ₹12850.00. Average gross returns during the bearing period accounted to ₹36000.00 per hectare. The pre-bearing cost for five years was ₹47145.00 with ₹6935.00, ₹7975.00, ₹9800.00 and ₹9950.00 for 2nd, 3rd, 4th and 5th year, respectively.

Gummagolmath *et al.* (2002) studied the cost of cultivation of mango in Dharwad district of Karnataka and revealed that per hectare total variable cost was ₹5840.91 in small, ₹6859.58 in medium and ₹6659.60 in large orchards. The low variable cost in small orchards was due to the absence of irrigation from 11 years onwards because all the small orchards carried pot irrigation upto 10 years. After 10 years irrigation was not necessary. In medium orchards, total maintenance costs were high due to intensive operations such as earthing up and weeding. Despite applying more

amounts of farmyard manure and plant protection chemicals, the large orchards incurred less cost due to operation of economies of scale in various agronomic practices.

Ozkan *et al.* (2002) studied the costs and returns of citrus (orange, mandarin, and lemon) production in Antalya, Turkey, based on data for the 1999-2000 production season. The data were collected from 105 citrus growers. Production costs per hectare for oranges, mandarins, and lemons were estimated to be 4223.9 million TL, 3459.5 million TL, and 3700.5 million TL, respectively. Net profits per hectare were 2526.1 million TL for oranges, 388.5 million TL for mandarins, and 379.5 million TL for lemon. Among the citrus crops investigated, orange production had the highest relative advantage (1.60).

Mali *et al.* (2004) conducted the study on economics of production and marketing of banana in Jalgaon district of Maharashtra during the year 2000-01 and observed that the total cost of cultivation of banana was ₹120539.12 per hectare. The yield per hectare worked out to be ₹533.14 quintals with gross returns amounting to ₹214867.24. The per quintal cost of production were found to be ₹220.06. The per hectare net profit was ₹79640.11.

Nandal and Punia (2004) in their study on economics of major fruit crops in western zone of Haryana during 2002-03 found that the total cost of cultivation of guava increased from ₹9185.90 per acre in the first year to ₹14976.74 in the seventh year and onwards and the yield increased from 2.50 quintal per acre in the third year to 109.50 quintal per acre during the seventh year and onwards with gross to be ₹60225 per acre. The total cost of ber cultivation increased from ₹7400.14 per acre in the first year to ₹9399.49 in the seventh year onwards and gross returns ₹34347.22 with a production of 64.20 quintal per acre which was found increasing from 1.25 quintal per acre in the third year to 64.20 quintal per acre in the seventh year.

Gangwar *et al.* (2005) in their study on an economic evaluation of kinnow mandarin cultivation in Punjab during 2002-03 year found that the total cost of cultivation per hectare of kinnow ranged from ₹28407.00 in the first year to ₹24226.00 in the fifth year and therefore, the total establishment cost of orchard was worked out to be ₹119107.00. The maintenance cost from the sixth year onward varied from ₹26157.00 to ₹43354.00 per hectare and the average gross returns per hectare amounted to ₹73672.00 per year.

Radha *et al.* (2006) revealed that the establishment cost in case of production of grapes amounted to ₹316174 per hectare. They also found that the various items under total production cost (₹176503 per hectare), manures and fertilizers (₹20768 per hectare) and the rental value of owned land (₹92956 per hectare) among fixed costs contributed a major share of 11.76 and 52.66, per cent respectively.

Sharma *et al.* (2006) conducted a study during 2002-03 to examine the economic feasibility of mango cultivation in Yamunagar district of Haryana by estimating the initial cost of establishing a mango orchard, amortization capacity of the orchard and profitability of mango cultivation vis-à-vis other annual crops. Results showed that mango growers incurred losses during the first four years of establishment of

mango orchards, the profit earned increased from the fifth year (₹3873 per year) to the tenth year (₹47477 per year) and thereafter became almost stagnant throughout the expected life of 30 years.

Kumar *et al.* (2007) revealed that per hectare production cost of apple in Himachal Pradesh on marginal, small, medium, semi-medium and large orchards was ₹131976, ₹127182, ₹127321, ₹128099 and ₹135149, respectively, while as net returns were highest on marginal orchards (₹153408) followed by large orchards (₹140049) and least for medium category orchards (₹129143).

Wagale *et al.* (2007) revealed that per hectare cost of cultivation of mango was ₹43198.00, ₹44310.00 and ₹48103.00 in small, medium and large orchards, respectively. The net return was ₹13594.00, ₹16747.00 and ₹18879 in small, medium and large orchards, respectively. Further, the cost benefit ratio at total cost of production was 1.26, 1.31 and 1.33 in small, medium and large farms, respectively.

Singh and Sayeed (2008) estimated that per hectare cost of aonla on an average was ₹28958.64, which was lowest at ₹25705.04 on marginal, ₹27388.08 on small, ₹29491.16 on medium and highest was ₹33250.04 on large farms. The major cost involved in the cultivation of aonla was reported on human labour followed by manure and fertilizer, plant protection and irrigation charges. On an average cost A, cost B and cost C were observed at ₹15989, ₹19562 and ₹22937, respectively and all the three costs increased with the increase in the farm size. The farmers above 4 hectares land managed their resources properly to raise the level of output.

Iqbal (2009) in his study on investment appraisal of mango and ber fruit production in Jammu district of J and K state during the year 2008 concluded that the per acre first year establishment costs of mango and ber were ₹5517.42 and ₹3869.89, respectively. Costs and returns of mango and ber were analysed by applying the cost concepts of cost A, cost B and cost C. The results revealed that on an average, per acre cost A, cost B and cost C in mango was ₹609.58, ₹1782.48 and ₹5164.32, respectively, whereas in ber they were ₹266.91, ₹1434.62 and ₹3917.12, respectively. The per acre per year average returns from mango and ber were worked out to be ₹7794.97 and ₹5769.11, respectively.

Yeware *et al.* (2010) conducted a study in Nanded district of Maharashtra state during the year 2007-08 to compare the costs, returns and profitability of Mrugbahar and Ambebahar sweet orange orchards. Costs and returns of sweet orange production were analysed by applying the cost concepts of Cost A, Cost B and Cost C with the help of tabular analysis. The results revealed that, on an average, Cost A, Cost B, and Cost C in Mrugbahar oranges was ₹32035.10, ₹55149.62, and ₹56197.90, respectively. In Ambebahar oranges they were ₹33951.30, ₹58473.31 and ₹60513.31, respectively.

On the whole, it could be concluded that cost of establishment of an orchard was high and major items of cost were preparation of land, digging and filling of pits, manuring, fertilization etc. The major items of cost of cultivation were human labour (family + hired), manures and fertilizers and rental value of land.

2.4. Review of Economic Viability Studies

Gupta and George (1974) studied the profit of Nagpur santra cultivation and

concluded that the project had a pay-back period of seven to nine years and yielded an internal rate of return of 29.3 to 45.9 per cent, depending upon the size of the orchard. The net present value and benefit-cost ratio even at high discount rate of 12 per cent varied from ₹4260.00 to ₹7910.00 per acre and 1.85 to 2.64, respectively according to the size of orchard. The productive life of an orange tree was found to be more than 24 years and optimum size of an orange orchard was between 1 to 2 acres.

Subrahmanyam (1986) examined the profitable lime cultivation in Andhra Pradesh and revealed that lime cultivation was profitable than the sweet orange with a pay-back period of four years, benefit-cost-ratio of 2.62 and a capital value of ₹36000.00 per hectare compares with six year pay-back period, 2.27 benefit-cost-ratio and ₹25602.00 per hectare capital value of sweet orange.

Thakur *et al.* (1986) in their study on the economics of kinnow cultivation in Himachal Pradesh revealed that the pay-back period in kinnow orchards was six years with average annual net returns at 10 per cent of ₹13,496.68 and benefit-cost-ratio at 10 per cent (over gross returns) was 3.04 and benefit-cost-ratio (over net returns) was 2.04. The internal rate of returns was 46.47 per cent.

Singh (1987) examined the feasibility of kinnow, mango and guava production in kandi area of Punjab. Cultivation of these fruits had been found to be financially feasible at 18 per cent of discounted rate. The benefit-cost-ratio had been found to be more than 1.6 and internal rate of return as high as 38.91, 46.4 and 48.4 per cent, respectively.

Subrahmanyam (1987) concluded that investment in mango cultivation in Karnataka gave internal rate of return around 30 per cent but also realized benefit-cost ratio above 2.00. The pay-back period was slightly more and it took 11 years from planting to recover the investment. The income from mango cultivation was comparatively low due to neglect of orchards because of price and yield fluctuations and prevailing method of marketing warranting early action to solve some of these problems.

Hugar *et al.* (1991) studied the economic feasibility of guava cultivation and observed that the net present value in guava orchards at 14 per cent discounted rate was ₹738042 per hectare. The benefit-cost-ratio was found to be 3.88 with internal rate of return (IRR) 57.82 per cent and pay-back period of 6 years.

Sikka *et al.* (1992) found that the pay-back period of apple cultivation in Himachal Pradesh was 16 years except Shimla where it was 15 years. The net present value at 12 per cent discount rate was ₹29872 per hectare in Shimla, ₹7627 in Mandi and ₹6365 per hectare in Kullu area. The overall net present value was found to be ₹14665 per hectare. The internal rate of return was 24.07 per cent, 17.84 per cent, 16.49 per cent and 20.43 per cent for Shimla, Mandi, Kullu and overall region, respectively. The benefit-cost-ratio was 1.34, 1.06, 1.09 and 1.16 for Shimla, Mandi, Kullu and overall region, respectively.

Singh and Khatkar (1994) studied the economic analysis of grape cultivation in Hissar and concluded that grape cultivation is an economically viable venture with internal rate of return 23 per cent and pay-back period of 8 years. The benefit-cost ratio was found to be of 1:8.

Gangwar and Singh (1998) in their study on Nagpur mandarin cultivation in Vidharbha region of Maharashtra during 1996-97 revealed that the net present value at 12 per cent discount rate varied from ₹40718 to ₹46564 depending upon the size of mandarin orchards. The internal rate of return ranged from 25.78 per cent to 28.83 per cent while as benefit-cost-ratio ranged from 1.38 to 1.45. The pay-back period was less than 8 years for all categories of Nagpur mandarin orchards.

Choubey and Atteri (2000) conducted the study on the economic evaluation of litchi production in Bihar during the year 1996-97. The benefit-cost-ratio and the net present value worked out at 10 per cent discount rate were found to be 1.4 and 21 per cent, respectively with pay-back period of 10 years. They also found that litchi was labour intensive fruit and more labour was required at the time of plantation.

Radha *et al.* (2000) studied the economic appraisal of citrus projects and revealed that the higher benefit-cost-ratio of 1.53:1 and the positive net present value of ₹40060.00 supported the fact that citrus project was economically feasible over long run of fifteen years. The internal rate of return calculated (26.03 per cent) was also higher than the discounted rate of interest (12 per cent).

Gangwar *et al.* (2005) examined the economic evaluation of kinnow mandarin cultivation in Punjab during the year 2002-03 and found that kinnow orchard was a profitable proposition, where the internal rate of return (IRR) at 12 per cent discount rate varied from 22.41 to 25.65 per cent, depending upon the size of orchards. The net present value and benefit-cost-ratio worked out to be ₹110803 and 1.435. The economic productive life of kinnow orchards was found to be 25 years.

Bakhsh *et al.* (2006) worked out benefit-cost-ratio and net present worth of growing mango orchard. Net present worth of ₹155607.16 per acre was estimated for the sampled respondents which indicated that mango cultivation fetched higher returns whereas benefit-cost-ratio was reasonably high and it came to be 2.61.

Radha *et al.* (2006) studied the economic analysis and production and marketing of grape in Andhra Pradesh and revealed that the pay-back period observed was 1.4 years, with benefit-cost-ratio of 1.45 and net present value at 15 per cent of ₹143804 and internal rate of return was found to be 49.45 per cent.

Sharma *et al.* (2006) in a study on mango cultivation in Yamunanagar district of Haryana during 2002-03 found that the net present value per hectare calculated at 12 per cent discount rate was ₹110165 during entire life of the orchard. Based on cost-benefit-ratio of 1:3 and internal rate of return of 25 per cent in mango orchards they concluded that mango cultivation in the district was profitable.

Iqbal (2009) during the year 2008 studied the investment appraisal of mango and ber in Jammu district and revealed that the pay-back period, benefit-cost-ratio, net present value per acre and internal rate of return of mango orchard was worked out to be 6.4 years, 1.51, ₹4802.52 and 18.85 per cent, respectively, whereas in case of ber it was found to be 6.7 years, 1.54, ₹3951.34 and 16.17 per cent, respectively.

Sharif *et al.* (2009) conducted a study at Social Sciences Institute, National Agricultural Research Centre, Islamabad during 2006 and analyzed that the Internal rate of return was found to be 33 per cent against current rate of interest on agricultural

loan with pay-back period of 7 years. Moreover, analysis indicates that 20 years is optimum economic life of citrus orchard, after that there is a declining trend of citrus production potential.

Supe *et al.* (2009) collected and analysed the data regarding cost of production, yield, prices and market availability of kagzi lime over a period of three years *i.e.* 2004-2006. It was concluded that growers must adopt micro-irrigation systems for this fruit crop as because it yielded approximately a net return of ₹58,000/ha/annum having benefit-cost-ratio of 2.88 and pay-back period of 2 years.

It can be concluded that the importance of net present value (NPV), pay-back period, internal rate of return (IRR) and benefit-cost-ratio (BCR) in the horticulture is very much important in assessing the economic viability of the orchard. These all depend upon the size of the orchards.

Chapter 3
Economic Theories of Marketing

3.1. Theory of Marketing

In empirical literature, numerous version of the definition of marketing are available. However, the definition from Kohls and Uhls (1980) seems most appropriate in our context. They defined marketing as "The series of activities involved in making available services and information which influence the desired level of production relative to market requirements and the movement of produce from the point of its production to the point of final consumption". Food and agricultural marketing means the movement of agricultural produce from the farm where it is produced to the consumer or manufacturer. Agricultural marketing also includes the marketing of agricultural inputs. It is important to recognize the overall role that marketing can play in a developing economy, for marketing per se can be regarded as a vital factor for development (Holton, 1953). In the past so many attempts have been made by the development economists to keep abreast the producers regarding the philosophy of markets and marketing systems but the credit goes to the father of Development Economics, Adam Smith (1976) who opened a new era in "The Wealth of Nations", by stating the theory of division of labour and thus inculcated the concept of specialization of human activities, the ability to mass produce and the need to exchange products and to develop the value of money mechanism to facilitate an exchange system. This concept has widened the range of economic activities in the form of processing, packing, storage, transportation, credit and advertisement etc.

3.2. Review of Marketing Channels, Costs and Price Spread Studies

Gill *et al.* (1980) studied the marketing pattern of citrus in Punjab state and revealed that on an average, contract price turned out to be ₹37.73 per quintal.

Regarding disposal of citrus fruit, 96.5 per cent of the total produce was sold by the contractors and 3.5 per cent by the producers themselves. The producers' share in consumers' rupee was the highest in local market whereas the producers' share in the distant market (Delhi) was less than 22 per cent.

Patel and Pawar (1980) studied the marketing of fruits in Mahatma Phule market, Bombay. The study showed that there was close relationship between wholesale prices and supply position, whereas there was no such pattern in case of retail prices. The producers' share in consumers' rupee was 33.07 per cent for sweet oranges. In case of apples, mangoes and grapes, the share of the producer in consumers' rupee varied according to the varieties and ranged from 33 to 54 per cent. The study also made some suggestions as standardization and grading, and establishment of cooperative marketing societies for getting better prices for the producers.

Auhurkar and Deole (1985) estimated the producers' share in the consumers' rupee for banana, sweet orange, mandarin orange and sour lime in the Marathwada region of Maharashtra and reported that it ranged between 28 per cent and 30 per cent. The margin of the intermediaries accounted for about 24 per cent of consumers' rupee while about 46 per cent to 48 per cent of the consumers' price went towards marketing costs.

Mondal (1986) conducted a study on 'Marketing of Pine apples in Meghalaya state' and analysis of price spread showed that net profit of the producer was highest (83.32 per cent) when the produce was sold directly to consumer. Net percentage margin received by the retailer when operating through 'producer-retailer-consumer' channel was 28.70 to 30.19 per cent, whereas, net share of wholesaler was 14.84 per cent to 15.20 per cent.

Singh and Sikka (1992) revealed that during marketing of apples in tribal areas of Himachal Pradesh, producers' share in consumers' rupee was 33.29 per cent. The marketing cost borne by orchardists was 41.07 per cent and margins of mashakhars and retailers were 2.23 per cent and 11.85 per cent of consumers' rupee, respectively.

Lepeha *et al.* (1993) studied the price spreads of Mandarin orange at Kalimpong wholesale and retail markets during the year 1990-91. Producers on an average got 57.14 per cent of the consumers' price. Out of 42.86 per cent of consumers' price, transport cost at producers' level was 4.29 per cent, packing cost and spoilage accounted for 0.71 per cent each and 0.71 per cent of the consumers' price went to the cost of loading and unloading. At the commission agent level, there was only a single cost, *i.e.*, cost of storage, and it was 0.71 per cent of the consumers' rupee and 7.15 per cent went as wholesaler's profit at Kalimpong retail market, transport cost accounted for 2.87 per cent of the consumers' rupee, loading and unloading 1.43 per cent, cost of helping hands 1.43 per cent and spoilage 1.43 per cent of the consumers' rupee. Retailer profit was 21.43 per cent.

Sidhu (1993) conducted a study to examine the price spread of kinnow through producers to pre-harvest contractors to retailer to consumers in Delhi, Ludhiana, Amritsar and local market. He revealed that about 70 per cent of kinnow was sold in Delhi market due to brisk demand and higher prices. The contractor did not prefer to

sell in the local markets like Malout, Abohar, etc. only a small quantity was sold in local Malout market.

Tewari (1994) examined the marketing of kinnow in Himachal Pradesh and observed that 60 per cent of kinnow growers preferred to sell their produce themselves. Producer → commission agent → wholesaler → retailer → consumer was the most popular channel as about 91 per cent of kinnow was marketed through this channel. Pathankot market happened to be the most popular among the kinnow growers of study area as about 97 per cent of kinnow was sent to this market alone.

Tomer *et al.* (1997) studied the marketing costs and margins for citrus (malta and kinnow) in Hisar and Sirsa districts of Haryana. The study revealed that producers' share in consumers' rupee was around 50 per cent when the producers directly sold citrus in the market, however, when sold through pre-harvest contractor, the share declined to about 40 per cent. The marketing margins charged by the middleman for citrus were higher which ranged from 14 to 18 per cent of the consumer price.

Ladaniya *et al.* (2003) examined the price spread of pomegranate. The study highlighted that pomegranate was mainly grown in Maharashtra, Karnataka and Rajasthan in the country. Over 90 per cent of produce was marketed in three production centres of Sangola, Malagaon, Rahuri through the following major channels: (1) Producer → Commission agent → Wholesaler → Retailer → Consumer, (2) Producer → Cooperative society → commission agent → Retailer → Consumer, (3) Producer → Commission agent (local) → Trader (distant) → Wholesaler → Retailer → Consumer. Local channels *viz.* (i) Producer → Consumer → (ii) Producer → Retailer → Consumer, (iii) Producer → Pre-harvest contractor → Retailer → Consumer were also followed through which 10 per cent of the produce was marketed. Producers with small holdings preferred cooperative society and marketing of produce through forming self-help groups for transportation. Packing, long distance transportation and commission charges accounted for 90 per cent of the marketing costs. Retailer's margin was in the range of 38.50 to 56.33 per cent in the price paid by the consumer.

Ladaniya *et al.* (2003) studied the marketing pattern of mosambi in selected districts of Maharashtra and observed that mosambi was marketed through three different channels such as (1) Farmers → Pre-harvest contractor → Commission agents → Retailer → Consumer, (2) Farmers → Commission agents → Wholesaler → Retailer → Consumer and (3) Farmers → Pre-harvest contractor → Commission agents → Buyers of distant market → Wholesaler → Retailer → Consumer. Commission and transportation charges mainly contributed towards cost of marketing incurred by producer. The maximum increase in the retail price of the mosambi was due to retailer's high margin.

Ajani (2005) estimated the marketing margins, net margins and profitability at different levels of marketing. Results revealed that tomatoes had highest marketing margin of 39.90 per cent and 41 per cent at both the wholesale and retail levels in the three markets. This implied a wide gap in prices between wholesalers and retailers. The study also revealed that fruits were cheapest in Oyingbo market. The least marketing margin was recorded in the sales of bananas at the wholesale level (19 per cent) and oranges (18.09 per cent) at the retail level. This implied a higher marketing

efficiency in some fruits (banana) than another (tomatoes) because a higher percentage marketing margin shows lower marketing efficiency.

Ladaniya *et al.* (2005) examined the channels of marketing, price spread and marketing efficiency, pattern of arrivals and prices; compared the economics of fresh grape marketing and estimated losses of fresh grapes at farm, wholesale and retail level in selected areas and markets in Maharashtra, India. The marketing channels identified were: (1) grower-commission agent-retailer-consumer for local and distant markets; and (2) grower-commission agent-trader-retailer-consumer for distant upcountry markets. Marketing efficiency was 1.27 when produce was directly sold by producers to retailers. When produce was sold to traders in vineyard, efficiency was 0.50, with producers' share of 33.65 per cent in consumers' rupee. In longer channels, efficiency was 0.48, with producers' share of 32.50 per cent in consumers' rupee. Raisin making was profitable and earned ₹50500.00 in addition to marketing grapes. Marketing of raisins was through commission agents. Losses in grapes were 1-1.25 per cent at farm, 5.5-8.65 per cent at wholesale and 12.25-16 per cent at retail level.

Shah *et al.* (2010) examined the marketing margins in citrus fruit business in Haripur district of NWFP, Pakistan during. The present study was conducted during the citrus production season of year 2008. The results revealed that growers received only 35.7 per cent of the final price, while rest of the 64.0 per cent price was accumulated into the profit basket of market functionaries. The share of various intermediaries was; contractors got 21.3 per cent, 9.8 per cent of the profit margin was taken away by the commission agents and 14.9 per cent by the wholesalers, while remaining 18.4 per cent profit margin went to the retailers and other functionaries. This indicated that the marketing system in the project area was not properly regularized and because of which the growers did not receive logical return for their crops.

On the whole, it could be concluded that the producers' share in consumers' rupee was relatively very low except in case of that channel where produce was sold directly by the producer to consumer. Price spread analysis revealed that, various market intermediaries were the highest beneficiaries in the marketing channels.

3.3. Review of Price Behaviour Studies

Vitonde *et al.* (1991) studied the marketing of mandarin oranges in Nagpur district of Maharashtra. They worked out the indices on the basis of prices during the period of 10 years *i.e.*, from 1976-77 to 1985-86. They had seen that the prices were high in the month of December, January and April as compared to other months and that was due to reduction in supply of oranges in relation to demand. The lower prices existed in October and May due to low quality of produce.

Autkar *et al.* (1994) examined the trends in seasonal fluctuation in arrivals and prices of kagzi lime in eight different markets of Akola and Buldana districts of Maharashtra during 1989-90. It had been observed that when major portion of produce reached the Akola market during peak period, the prices generally ruled low which resulted in decrease in farmer's income. They also studied that there was a sudden rise in the prices of kagzi lime from month of January (₹4048.50 per tonne) to April

(₹4427.00 per tonne) and then from the month of May the prices fell down. On the contrary, the arrivals were found to be less from January (317.20 tonnes) to April (304.80 tonnes) and from May onwards the arrivals started increasing and were highest in the month of November (2579.30 tonnes).

Bagde *et al.* (1996) studied the data pertaining to monthly prices of apple in Nagpur. It was observed that monthly prices in respect of apple varied in the range of ₹598.00 to ₹925.00 per quintal. The lowest prices were recorded in the month of August while the highest in the month of March. Out of nine months, four months had shown the prices less than the annual average prices of apples when compared to the arrivals. It was observed when arrivals were less the prices tend to increase gradually. The average arrivals in case of apple were 3423 quintals while the average prices of apples observed to be as 763.33 per quintal. The arrivals of apple were noticed throughout the year except April, May and June. The highest arrivals were noticed in the month of September, October, while lowest was noticed in the month of January, February, March and July. 75 per cent of arrivals of apple were observed during six months from July to December.

Chavan (2004) studied the marketing of pomegranate in Sangli district of Maharashtra and revealed that the high indices for arrival were during the month of July to January and highest in August *i.e.*, 189.37. In case of prices, the high indices were seen during July to January and highest in November *i.e.*, 109.27. At the overall level the variability in arrivals of pomegranate were 71.81 per cent during last 10 years. The minimum variability was observed in the month of December *i.e.*, 26.45 per cent and maximum in the month of May *i.e.*, 129.52 per cent.

Anwarul Haq *et al.* (2006) studied the data pertaining to seasonal variation in price of lemon in Bangladesh. It was observed that the price of lemon changed in respect of its availability in the market. The lowest price according to price indices prevail in June (74) due to peak harvesting period. The price starts rising from October and reached the peak level in February (186) due to limited supply in the market. The price indices ranged between 74 to 186.

The studies on price fluctuation have indicated a very high degree of seasonal fluctuations in prices with high prices when arrivals were low and vice-versa, which was attributed to seasonal behavior and perishable nature of the produce.

3.4. Review of Constraints in Production and Marketing Studies

Kadam (2000) conducted a study on constraints in marketing management of oranges faced by farmers was conducted in 1998-99 in Amravati district, Vidarbha region, Maharashtra, India. Results indicated that a majority of farmers had a medium level of marketing management constraints. Education, land holding, socio-economic status, management orientation, achievement motivation, mass media exposure and knowledge emerged as the important factors which affected marketing management. Major constraints reported by farmers were absence of pre-cooling centres, absence of cold storage centres, high transportation cost and lack of processing units.

Sharan and Singh (2002) examined the marketing problems faced by kinnow growers in Rajasthan. The study revealed that selling of produce through self marketing by growers was found profitable in comparison to contract sale to pre-harvest contractors. The major problems highlighted by the study were lack of support price, lack of growers organization, delay in payment, lack of marketing information, lack of cold storage facilities and lack of better and cheaper packing material. The growers also reported other problems like lower prices due to seasonal gluts, lack of stay arrangements in the market, malpractice in weighing, etc.

Thilagavath *et al.* (2002) examined the economic viability of dry land horticulture in rainfall vertisols (Southern Tamil Nadu) and the study revealed that fruit crops were gaining importance on farms in the southern zone of Tamil Nadu. Investment analysis indicated that farmers' practice of dryland horticulture was economically feasible. Fruit crops exhibited remunerative discounted cash flow measure compared with annual crops. Technology awareness, maintenance of orchard, marketing, non-availability of labour, rainfall, credit availability and social problems, however, were the major constraints.

Bhole (2004) studied the problems in marketing of oranges in Vidharbha. He indicated main problems such as delayed payment, breaking contract as orange prices slash, cut in payment of farmers in the event of loss of fruits due to dropping, high commission charges, high transportation, loading and unloading charges and delayed payment of commission agents.

Phuse *et al.* (2008) studied the various constraints faced by the Nagpur mandarin orange growers during mandarin production by selecting 200 growers from Amravati district of Vidharbha region. The data collected and analysed revealed that unavailability of labour and irregular electric supply were the major production constraints faced by the 100 per cent respondents. The respondents also reported about unavailability of quality seedlings (89.00 per cent), high rate of insecticide and fertilizer (84.00 per cent) and improper road facility (66.00 per cent). Regarding plant protection measure, most of the respondents (69.16 per cent) had problem of high cost of insecticides and fungicides. It was also felt by the growers that rate of interest was very high (96.00 per cent) along with labour charges (93.00 per cent). Lack of knowledge about various cultivation practices, poor economic status of the respondents and marketing constraints were also predominant. The 100 per cent grower experienced that there was no assurance to sale the produce in a market with minimum support price. There was 86 per cent orange grower who felt lack of market information as a major constraint. It was followed by lack of storage facility (83.50 per cent), high cost of transportation (77.00 per cent) and lack of group marketing (72.50 per cent). The farmers were also facing the problem of low market rate due to middlemen.

Ghafoor *et al.* (2010) studied the marketing problems faced by kinnow growers of tehsil Toba Tek Singh (Pakistan). The study revealed that lack of storage facilities and non-availability of cartons appeared to be the major problems faced by kinnow growers. Late payments by the dealers, less price of kinnow in markets, monopoly of middle men, packing and loading, high carriage and other handling charges, also perceived to be very important factors contributing towards marketing problems of kinnow.

From the above, we can conclude that major production and marketing constraints associated with the fruits were high cost of inputs, lack of skilled labour, lack of storage facilities, unavailability of quality seedlings, large numbers of middlemen, unregulated markets, high commission charges, high cost of transportation and lack of market information.

Chapter 4

Production and Marketing of Citrus in the Jammu Region: A Case Study

4.1. Introduction

The term citrus fruit includes different types of fruits and products. Although oranges are the major fruit in the citrus fruits group, accounting for about 70 per cent of citrus output, the group also includes small citrus fruits (such as tangerines, mandarines, clementines and satsumas), lemons and limes and grapefruits. The leading processed form in the group is orange juice. Citrus fruits are produced in many countries around the world, although production shows geographical concentration in certain areas. Mediterranean countries are the leading area as producers for the international fresh market and Spain plays a dominant role in the area. Most orange juice production is concentrated in only two areas, Sao Paulo in Brazil and Florida in United States of America (Brazil is by far the largest orange juice exporter).

Citrus fruits are among the important commercial fruit crops and evaluating their economics of production and marketing will help the fruit growers of this region to a greater extent as how to make their cultivation and marketing more profitable besides will also act as a guideline for the planning of policy planners/scientists. Keeping in view the importance of the citrus fruit for our orchards, and the facts described, a research study entitled, *"Economics of Production and Marketing of Citrus in Jammu region of Jammu and Kashmir State"*, has been undertaken.

4.2. Importance of the Study

The economic aspects of fruit cultivation are not less important as well maintained and established orchards give better returns than field crops from the same piece of land. It may also be mentioned that there are many factors which may enhance the production of citrus fruits but among them cost and return coupled with marketing are considered to be the key factors for increasing the production. The growers before prioritising the preferences for establishing orchard ensures its cost and return factor, which is the main motivation factor for bringing more area vis-à-vis giving lot of attention. Therefore, it also becomes imperative to workout its payback period which helps to various financial institutions and policy planners to provide the credit support. Moreover, alongwith the production, the role of marketing opportunities of fruits is equally important, as the farmers can ensure the reasonable return for their produce and also a legitimate share in the price paid by the consumers.

Although area under citrus and its production has shown a steady increase over time, yet their marketing aspects (comprising of marketing cost, marketing margin, marketing loss and price spread) has all along been almost neglected and at present marketing facilities for citrus are inadequate. Under the existing marketing practice, before the produce reaches to the end user, it has to be handled and passed through a long chain of various intermediaries, with the result that the producers are getting a small share of consumers' rupee. Therefore, working out the price spread particularly in citrus fruits provides an opportunity to know the difference between the price received by the orchardists and price paid by the consumer which comprises cost of undertaking and rendering market services such as assembling, grading, transporting, processing, wholesaling, retailing and the margins of the intermediaries, charges, sale tax etc, as they are too wide because to its perishable nature, seasonality of production, spatial distribution of citrus plantation far off from consuming centres, inadequate cold storage and credit facilities and lack of comprehensive marketing information. All these forces compel the growers to sell their produce unprocessed and immediately after harvest, resulting gluts in the markets and thereby fall in prices and hence lower returns.

4.2.1. Objectives of the Study

1. To estimate resource use efficiencies in citrus cultivation
2. To study the costs and returns
3. To study marketing channels, marketing costs and price spread
4. To study trends in arrivals and prices in regulated markets of Jammu
5. To identify the constraints in production and marketing

4.2.2. Hypotheses or Assumptions

- ☆ Citrus is a profitable commercial crop.
- ☆ There exists a greater scope for rational allocation of farm resources on citrus orchard.
- ☆ The cost and return structure vary with area of citrus orchard.

☆ Producer has maximum share in consumer rupee when sold through direct marketing channel whereas in other channels, a considerable proportion of consumer rupee is appropriated by market intermediaries.

4.2.3. Scope and Limitation of the Study

4.2.3.1. Scope

☆ Study on resource structure is important for implementation of policy for citrus cultivation.

☆ Input-output relationship and resource use efficiency will show the positive or negative significance of different independent variables on dependent variable.

☆ Analysis on costs and returns of citrus production will generate useful information regarding the different items of costs involves in citrus production and net returns after meeting all expenses.

☆ Study of marketing cost components and price spread in different marketing channels will help to identify the producers' share in consumers' rupee and also most efficient channel.

☆ The study would indicate the possible measures to make the citrus production most profitable and marketing practices most effective.

4.2.3.2. Limitation

☆ The study was based on sample survey of selected citrus orchardists from major citrus growing area of Jammu region. The information was collected by personal interview method based on respondents' remembrance, past experience and his involvement. Though maximum possible efforts were made to collect reliable information, yet there may be some lacunae in the information given by respondents.

☆ Owing to time and resource constraints to research scholar, a limited size of sample (192 orchardists) was taken. A larger sample size would definitely tend to improve the reliability of the estimated co-efficients.

☆ The study was conducted in a selected area, therefore, the results may not be applicable to each area of citrus cultivation but it may be helpful as a guideline for further studies.

4.3. Methodology

The case study "*Economics of Production and Marketing of Citrus in Jammu region of Jammu and Kashmir State*" was carried out during the year 2009. The sampling structure, experimental procedures and techniques adopted during the course of investigation have been described in this chapter.

4.3.1. Locale of Study

The present study was conducted in Jammu region of Jammu and Kashmir state. Jammu, Rajouri, Kathua and Samba districts were purposely selected on the basis of

highest area under citrus fruit crop. The study was confined to the three blocks from each district for citrus fruit crop.

4.3.2. Climate and Rainfall

The Jammu and Kashmir state is situated between 32°.17′–36°.58′ North latitude and 37°.26′–80°.30′ East longitude. The state is the northern most part of the India is girdled by Tibet to the east, China and Afghanistan to the north, to its west is Pakistan. To its south lies the states of Punjab and Himachal Pradesh and have been divided into three agro climatic divisions' *viz.* outer plain and outer hills, middle mountains and Kashmir valley and inner Himalayas (Ladakh). The outer plains and outer hills includes Kathua and Jammu districts, extends up to Shivalik hills in the north. According to NARP (National Agricultural Research Project) classification this state has been divided into four agro-climatic zones: Temperate, Sub-tropical, Intermediate and Cold-Arid zone. In summer hot dry winds from plains of Punjab make it very hot, dust storms are common with occasional rain. Rainfall occurs from July to September. Average rainfall is 1143 mm and temperature shoots up to 46°C during May to June. During December to February temperature is between 13.5°C to 20°C, rainfall is 150 mm and in upper reaches there is snow. In the region of middle mountains and Kashmir valley winters are very cold. There is snowfall during winter and sub-zero temperature. In valley, winters are long and summers are short but pleasant. Average rainfall is 732 to 854 mm. The maximum temperature hardly ever goes beyond 35°C. In region of inner Himalayas, days are hot and there are no clouds in the sky. Winters are very cold. Temperature falls as low as minus 23°C. The average rainfall is 976 mm and humidity level is very low.

The four districts which were purposely selected for the present study were Jammu, Rajouri, Kathua and Samba of Jammu region. Jammu region which falls under sub-tropical zone as per NARP classification has hot and dry climate in summer and cold climate in winter. The climate of Rajouri district of the Jammu region varies from semi-tropical in the southern part to temperate in the mountainous northern part. The sub-tropical region receives regular monsoons whereas the northern part prone to hailstorms experiences excessive rains. The mean temperature of the Samba, Kathua and Jammu district of the Jammu region varies between the minimum of 3°–4°C from December to January to the maximum of 43°–47°C from May to June. The rainy season of the area normally starts from the end of June or in the first fortnight of July.

4.3.3. Collection of Data

The primary data from growers of citrus was collected by survey method, using well-designed schedule (Table 4.1) which consists of XIX parts. Collection of data was done by the personal interview method. The schedule was pre-tested before using for actual data collection. In addition to this collection of information regarding marketing of the citrus was done by visiting growers, various markets and contacting the different intermediaries involved in marketing of citrus. Also to study the various services provided and charges of varied market functionaries, commission agent/ forwarding agent, wholesalers and retailers were selected randomly and the required information as well as data was obtained from them.

Table 4.1: Schedule

I. Information of the Fruit grower and his/her family

1.	Name of the Respondent	:	
2.	Age and Sex	:	
3.	Marital Status	:	Married/Un-married/Widow/Widower
5.	Main Occupation	:	
6.	Village	:	
7.	Block	:	
8.	District	:	
9.	Family Status	:	

Sl.No.	*Name of the family member*	*Age*	*Sex*	*Education*			
				I	*P*	*HS*	*G*

I: Illiterate; P: Primary;' HS: High School; G: Graduate.

II. Family income

Sl.No.	*Farmer*	*Farmers Engaged in*			*Income per Year*			*Total*
		Farm	*Business*	*Service*	*Farm*	*Business*	*Service*	
1.	Adult							
	Male							
	Female							
2.	Children							

III. Land use pattern (2009-10)

Survey No.	Type of Soil	Total Area	Waste Area (ha) Permanent Waste	Area under Cultivation	Land Leased in	Land Leased Out	Source of Irrigation	Land Revenue	Rent Paid (Rs.)	Rent Received	Per ha Market Value Dry-Irri.	Per ha Govt. Value of Land

IV. Cropping pattern (2009-10)

Survey No.	Cultivable Area (ha)	Kharif			Rabi			Zaid			Perennial			Total Cultivated Area (ha)
		Crop	Variety	Area Dry-Irri.	Crop	Variety	Area Dry-Irri.	Crop	Variety	Area Dry-Irri.	Crop	Variety	Area Dry-Irri.	

V. Inventory

Sl.No	Kind	No.	Area (m^2)	Type of Construction	Year of Construction	Cost of Construction	Expected Life Period (Yr.)	Present Value	Share	Repair
A.	**Building**									
	1. Residential house									
	2. Farm house									
	3. Store house									
B.	**Irrigation structure**									
	1. Well									
	2. Tube well									
	3. Others									

VI. Livestock

Sl.No.	Kind of Animal	No.	Breed	Age (in years)	Owned/ Purchased	Month and Year of Purchase	Purchase Value (Rs.)	Present Value	Expected Life Period (Yrs.)	Remarks
1.	Bullock									
2.	Cow									
	(a) Local									
	(b) Hybrid									
3.	Buffalo									
4.	Sheep, goat, poultry and other									

VII. Implements and Machinery

Sl.No.	Item	No.	Year of Purchase	Purchasing Price (Rs.)	Expected Life Period (Yrs.)	Present Value	Repairs during the Year (Rs.)
(a)	Machinery						
	1. Tractor						
	2. Oil engine (H.P.)						
	3. Electric motor (HP)						
	4. Sprayer/duster						
	5. Others						
(b)	Ag. Implements						
	1. Ridger						
	2. Iron plough (Big)						
	3. Iron plough (Small)						
	4. Clod crusher						
	5. Seed drill						
	6. Harrow						
	7. Hoe						
	8. Bullock cart						
	9. Others						
(c)	Tools						
	1. Ghameli						
	2. Pickaxes						
	3. Spade						
	4. Weeding hooks						
	5. Sickle						
	6. Ropes						
	7. PVC crates						
	8. Others						

VIII. Labour

	Working Numbers			No. of Months Employed						Payment Made 1. Cash 2. Kind
	M	W	C	Kharif @			Rabi @			
				M	W	C	M	W	C	
Family Labour										
Casual Labour										
Permanently hired labour										

IX. Inter Crops taken in Orchard

Sl.No.	Season and Name of Crop	Dry/Irrigated	Area	Cost of Cultivation	Total Income	Benefit or Loss
1.	Kharif					
2.	Rabi					
3.	Zaid					
4.	Other fruit crops					

X. Establishment Cost or Initial Overall Cost for Citrus Orchard

1. Year of plantation
2. Variety
3. Area
4. Number of plants
5. Planting distance

Sl.No.	Operations	Material Used			Labourers		Hired Labour Charges	Machinery Charges (Rs.)	Total Value (Rs.)
		Type	Qty.	Value	Family	Hired			
1.	Layout								
	(a) Preparatory tillage								
	(b) Digging of Pits								
	(c) Filling of Pits								
2.	Manuring								
3.	Fertilizer application								
4.	Plantation material								
5.	Cutting								
6.	Irrigation								
7.	Training and pruning								
8.	Plant protection								
9.	Interest on Working capital								
10.	Other operations								

XI. Orchard Establishment Costs

Item	Quantity	Price/ Quantity	Operational Costs in Terms of		
			Animal Hours	Tractor Hours	Human Hours
			and cost per hour		
Present age of plant					
Pre sowing irrigation					
Land preparation					
— No. of ploughing					
— No. of plankings					
— Leveling of land					
— Others					
Planting charges					
Sprayer and implements					
Fencing					
Intercropping					
F.Y.M. application					
Year 1.					
2.					
3.					
4.					
5.					
6.					
7.					
Fertilizer application					
a) __________					
Year 1.					
2.					
3.					
4.					
5.					
6.					
7.					
b) __________					
Year 1.					
2.					
3.					
4.					
5.					
6.					
7.					

Contd...

XI–*Contd...*

Item	*Quantity*	*Price/ Quantity*	*Operational Costs in Terms of*		
			Animal Hours	*Tractor Hours*	*Human Hours*
			and cost per hour		
c) ________					
Year 1.					
2.					
3.					
4.					
5.					
6.					
7.					
Irrigation charges					
Year 1.					
2.					
3.					
4.					
5.					
6.					
7.					
Plant Protection					
Name of insecticides/pesticides					
Year 1.					
2.					
3.					
4.					
5.					
6.					
7.					
Pruning, Training and Weeding					
Year 1.					
2.					
3.					
4.					
5.					
6.					
7.					
Any other precaution taken					
Year 1.					
2.					
3.					
4.					
5.					
6.					
7.					
Expected age of the plant: ________________					

XII. Cost of irrigation

	Fuel Required/Hr	*Rate/Unit Fuel*	*Total Amount*	*Area Covered*
Electric motor				
Diesel engine				
Tractor				

XIII. Orchard Maintenance Costs

Item	*Quantity*	*Price/ Quantity*	*Operational Costs in Terms of*		
			Animal Hours	*Tractor Hours*	*Human Hours*
			and cost per hour		
F.Y.M. application					
Year 8.					
9.					
10.					
11.					
12.					
13.					
14.					
15. 15 onwards.					
Fertilizer application					
a) ________					
Year 8.					
9.					
10.					
11.					
12.					
13.					
14.					
15. 15 onwards					
b) ________					
Year 8.					
9.					
10.					
11.					
12.					
13.					
14.					
15. 15 onwards.					

XIII–*Contd...*

Item	Quantity	Price/ Quantity	Operational Costs in Terms of		
			Animal Hours	Tractor Hours	Human Hours
			and cost per hour		
c) __________					
Year 8.					
9.					
10.					
11.					
12.					
13.					
14.					
15. 15 onwards.					
Plant Protection					
Name of insecticides/pesticides					
Year 8.					
9.					
10.					
11.					
12.					
13.					
14.					
15. 15 onwards.					
Pruning, Training and Weeding					
Year 8.					
9.					
10.					
11.					
12.					
13.					
14.					
15. 15 onwards.					
Any other precaution taken					
Year 8.					
9.					
10.					
11.					
12.					
13.					
14.					
15. 15 onwards.					

XIV. Year-wise Production of Different Fruit Crops

Sl.No.	*Name of Fruit*	*Year of Plantation*	*Production Quantity with Year of Bearing*	*Price/ Qt.*
1.	Orange		Year 1.	
			2.	
			3.	
			4.	
			5.	
			6.	
			7.	
			8.	
			9.	
			10.	
			11.	
			12.	
			13.	
			14.	
			15. 15 years onwards	
2.	Kinnow		Year 1.	
			2.	
			3.	
			4.	
			5.	
			6.	
			7.	
			8.	
			9.	
			10.	
			11.	
			12.	
			13.	
			14.	
			15. 15 years onwards	
3.	Lemon		Year 1.	
			2.	
			3.	
			4.	
			5.	
			6.	

Contd...

XIV–*Contd...*

Sl.No.	Name of Fruit	Year of Plantation	Production Quantity with Year of Bearing	Price/ Qt.
			7.	
			8.	
			9.	
			10.	
			11.	
			12.	
			13.	
			14.	
			15. 15 years onwards	

XV. Production and disposal

Total Product in qtls.	Losses Due to Pest, Diseases	Use for Home Consumption	Given on	Marketed

XVI. Borrowing from Financial Resources

Sl.No.	Particulars	Cooperatives		Commercial Banks					Govt.	Total
1.	Amount taken	State	Central	SBI	PNB	J&K	RRB	Other		
2.	Period of repayment									
3.	Type of credit									
4.	Rate of interest									

XIX–Contd...

2. **Marketing problems**
   (*a*) Not getting remunerative price for the produce
   (*b*) Packing material is costly
   (*c*) Packages are not returned to the growers
   (*d*) Less demand of fruits because of competition of other fruits
   (*e*) Cheating by middlemen
   (*f*) High cost of transportation
   (*g*) High commission charges
   (*h*) Non receipt of payment in time
   (*i*) Lack of market information
   (*j*) Un-organised marketing and low price paid to farmers
   (*k*) High perishability of the fruits
   (*l*) Non-availability of market
   (*m*) Lack of processing units and co-operative societies

Secondary data was collected from various published sources such as bulletins of the Ministry of Horticulture, Govt. of India, Directorate of Economics and Statistics, Govt. of India, Directorate of Economics and Statistics, Govt. of Jammu and Kashmir state and Directorate of Horticulture Planning and Marketing, Govt. of Jammu and Kashmir state.

4.3.4. Sampling Structure

A multi stage sampling was adopted for the selection of samples, with districts, blocks, villages and orchardists as the first, second, third and fourth stage sampling units for citrus crop. Rajouri, Kathua, Jammu and Samba districts were selected because these four districts covered the maximum area under its cultivation (Rajouri covered 22.93 per cent, Kathua 20.56 per cent, Jammu 16.37 per cent and Samba 12.67 per cent out of the total area under citrus cultivation in Jammu region). Among the citrus fruits, orange, kinnow and lemon were selected. Then three blocks from each district were selected on the basis of area under citrus fruit cultivation and from each block two villages were selected. The ultimate units, that is, orchardists were selected randomly from each village so as to constitute a total sample of 192 (8 from each village) orchardists from the whole area under study.

4.3.4.1. List of Selected Districts

Four districts for citrus fruit falling in Jammu region were selected for this study:

Selected Districts

1. Jammu
2. Rajouri
3. Kathua
4. Samba

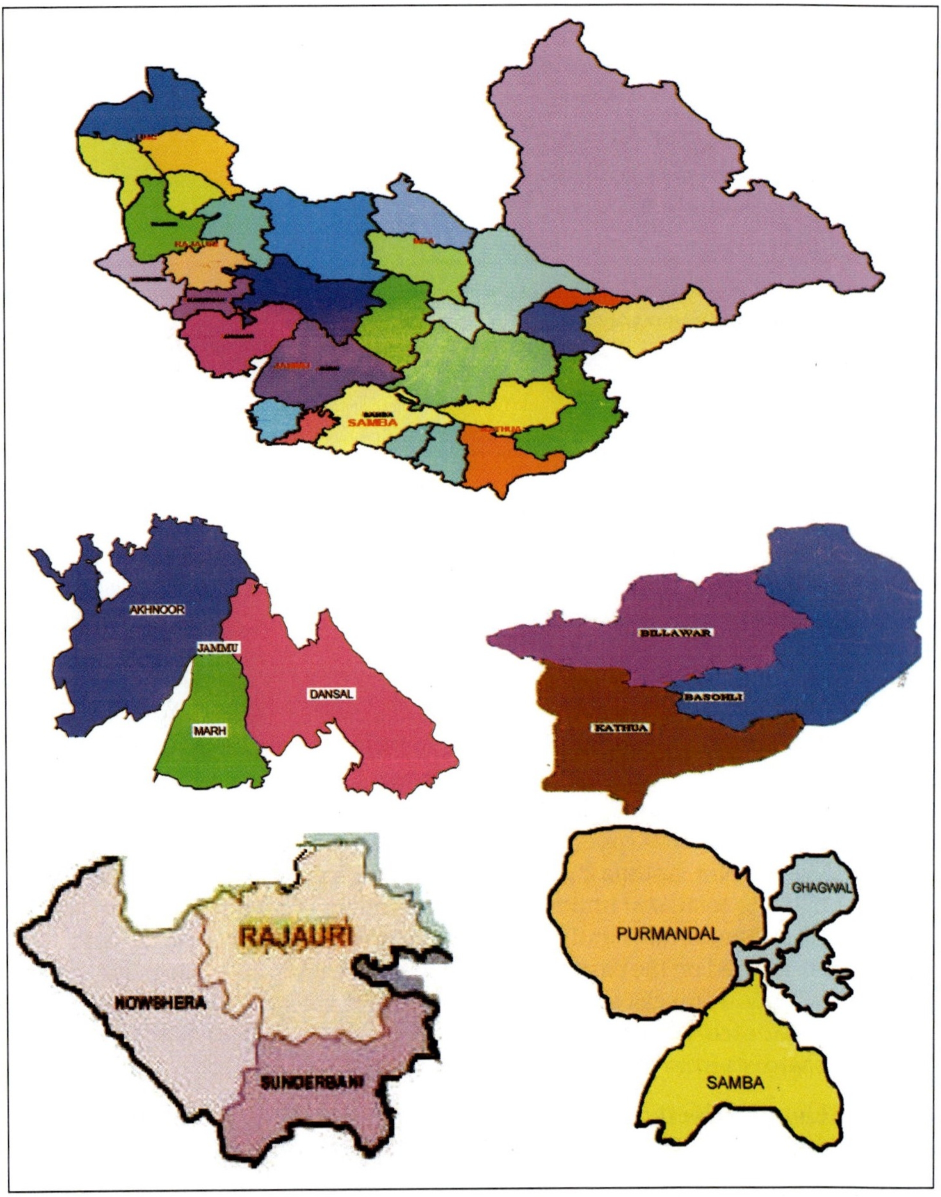

Figure 4.1: Map of Study Area

4.3.4.2. Selection of Blocks and Villages

The details of the selected blocks and villages are represented in Table 4.2.

Table 4.2: Details of Selected Blocks and Villages of the Study Area

Fruit	*District*	*Block*	*Village*	*Variety*	*Area(acres)*
Citrus	Jammu	Marh	Naibasti	Orange	7.50
				Kinnow	2.50
			Panjore	Kinnow	10.00
		Akhnoor	Nara	Orange	10.58
			Mandoh Fandwal	Orange	10.63
				Kinnow	1.50
		Dansal	Chak Rakhwala	Orange	7.75
			Dok Wajrian	Kinnow	14.25
	Rajouri	Rajouri	Chatiyar	Orange	7.90
			Dangri	Kinnow	10.25
		Sunderbani	Bajabain	Kinnow	24.75
			Bajwal	Orange	29.26
		Nowshera	Mayal Devi	Kinnow	27.13
			Seri	Kinnow	7.76
	Kathua	Kathua	Sahar	Lemon	13.75
			Pranta	Kinnow	4.50
				Lemon	4.75
		Basohli	Plahi	Kinnow	8.50
				Orange	2.50
			Sandhar	Kinnow	9.675
				Lemon	4.15
		Billawer	Dewal	Orange	6.175
			Makwal	Kinnow	17.63
	Samba	Purmandal	Thalori	Orange	18.26
				Kinnow	8.50
			Utterbani	Orange	13.75
				Kinnow	2.50
		Ghagwal	Ghagwal	Kinnow	17.50
			Rattanpur Sarara	Kinnow	19.75
		Samba	Amli	Kinnow	6.25
			Nadh	Kinnow	14.50

The data was collected on various aspects of established orchards, maintenance cost and economic returns from the orchards from 192 orchardists (8 from each village) who were randomly selected. The age of these orchards ranged from less than one year to 28 years.

From each selected village, the citrus orchardists were grouped into following four size groups of holdings

1. Marginal : 0.01 to 2.50 acres
2. Small : 2.51 to 5.00 acres

3. Medium : 5.01 to 7.50 acres
4. Large : 7.51 acres and above

But due to small sample size of lemon orchardists (15 orchardists only) the sample could not be divided into different size of holdings that is why the cost and returns, economic viability of lemon was clubbed for each and every size of holding.

4.3.5. Quantification of the Variables

The various inputs used in the production of citrus along with various costs and returns concepts were quantified as follows:

4.3.5.1. Human Labour

It included both hired and family labour. Most of the labour force engaged in fruit production comprised of family labour. However, the cultivator had to engage hired labour also. Human labour cost comprised of wages actually paid to the hired labour as also those paid to the labour obtained on contract for the whole year or part thereof and imputed value of labour put in by the family members working on the orchard. Existing casual labour wages for different operations were used to work out the total wage bill of labour employed per hectare of any fruit.

4.3.5.2. Bullock Labour

Hired bullock labour charges were considered for 8 hours a day, actually paid in the locality. Family bullock labour charges accounted equal to charges paid to the hired bullock labour.

4.3.5.3. Tractor Charges

Tractor was used in the orchard generally at the establishment time for the leveling of the land. Tractor charges actually paid by the orchardists were calculated by multiplying the area of the land with the market rate charged per unit of the land.

4.3.5.4. Manures and Fertilizers

This item included the expenditure incurred on the purchase of chemical fertilizers and farmyard manure used for the production of fruits on the sample orchards. The farmyard manure used at the orchard was assessed at the prices prevailing in the study area. Similarly, the physical quantities of different fertilizers used were multiplied with the market price.

4.3.5.5. Irrigation Charges

The study area was mainly rainfed, so the irrigation was mainly done from the tube wells with the help of electric motors or diesel motors. In case of electric motors electricity bill actually paid by the orchardists was considered, while in case of diesel motors diesel utilized was considered. But there were some areas of Akhnoor, Rajouri and Kathua where crop was totally dependent on natural rain.

4.3.5.6. Plant Protection

The various plant protection chemicals used per hectare for the orchards were assessed. Charges actually paid by the orchardists were calculated by multiplying the quantity of the various pesticides, insecticides, fungicides and weedicides with the market rates charged per unit of each plant protection item used.

4.3.5.7. Planting Material

The material was mainly the plantings which was purchased by the orchardists or raised on their own. The market value of these plantings was considered. In some areas planting material was provided by the Department of Horticulture free of cost.

4.3.5.8. Land Revenue

The land revenue actually paid by the orchardists to the government was considered.

4.3.5.9. Depreciation

The depreciation on the farm buildings, farm implements and machinery was considered at the rate of 12 per cent per annum which is similar to Choubey and Atteri (2000).

4.3.5.10. Gross Returns

Gross returns from the fruit crops per hectare and per orchard were obtained by the value of the total fruit harvested during the year. These were the post harvest market prices in the study area.

4.3.5.11. Rental Value of Land

The rental value of land was calculated as the $1/6^{th}$ of the gross produce value excluding the land revenue.

4.3.5.12. Interest on Working Capital

Interest on working capital was charged at the rate of 12 per cent per annum for the period of six months. The working capital was worked out on the current expenses such as expenditure on human labour, bullock labour, tractor charges, planting material, manures and fertilizers, irrigation charges etc.

4.3.5.13. Interest on Fixed Capital

The value of total assets was calculated and then 12 per cent of average values of fixed assets were considered as the total interest on fixed assets.

4.3.6. Economic Analysis

4.3.6.1 Production Function Analysis

In order to study the relationship between output and various inputs used, Cobb-Douglas production function was used. This function is used extensively in agricultural production function analysis. The functional form applies is written as

$$Y_t = \beta_0 \left(\prod_{i=1}^{n} X_i^{\beta_i} \right) u_t \quad (i = 1, 2, 3, ..n) \tag{4.1}$$

where,

Y and X_i (i =1, 2, 3, ..n) are the output and levels of inputs. The constant β_0 and β_i's (i = 1, 2, 3, ..n) represent the efficiency parameters and the production elasticities of the respective input variables for the given population at a particular period, t.

The fitted Cobb-Douglas production may be written for the present case with five input variables as:

$$Y = a_0 x_1^{b1} x_2^{b2} x_3^{b3} x_4^{b4} x_5^{b5} \quad (4.2)$$

On log transformation, the above function can be transformed to a linear form as:

$$\text{Log } y = \log a_0 + b_1 \log x_1 + b_2 \log x_2 + b_3 \log x_3 + b_4 \log x_4 + b_5 \log x_5$$

$$\text{Or } \log y = \log a_0 + b_i \sum_{i=1}^{5} \log x_i \quad (4.3)$$

where,

Y: Output (q/acre)

x_1: Human labour (man days per acre)

x_2: Manure + fertilizers (kg/acre)

x_3: Expenditure on plant protection (₹/acre)

x_4: Expenditure on irrigation (₹/acre)

x_5: Expenditure on training and pruning (₹/acre)

a_0: Constant

b's: Elasticities of production of respective resource categories

To examine the productivity of different inputs used in production of studied fruits, marginal value productivities of inputs were estimated at geometric mean levels of inputs. To calculate Marginal Value Productivity (MVP) of resource x_i, the following formula was used.

$$\text{MVP} = \hat{b}_i \frac{GM(Y)}{GM(x_i)} \times P_y \quad (4.4)$$

where,

MVP (x_i): Marginal value productivity of ith resource

$\hat{b}_i$: Regression coefficient (estimated)

GM (Y): Geometric mean of output

GM (x_i): Geometric mean of inputs

P_y: Price of output

4.3.6.2. Cost and Returns

The total input costs of citrus production was distributed under three heads using the cost concepts A, B and C.

Cost A included the cost on hired human labour, total bullock labour, planting material, value of manures and fertilizers, irrigation charges, depreciation on implements and machinery, land revenue, interest on working capital and establishment cost of citrus orchard.

Cost B represented cost A plus the imputed cost on account of rental value of the land and interest on fixed capital.

Cost C comprised of cost B plus imputed value of family labour. Thus cost C represented the total cost of cultivation.

4.3.6.3. Economic Viability

The economic viability were assessed using net present value (NPV), pay-back period, internal rate of return (IRR) and benefit–cost-ratio (BCR).

4.3.7. Net Present Value

Net present value (NPV) of an investment is the discounted value of all cash inflows and cash outflow of the project during its life time. It can be computed as

$$NPV = \sum_{t=0}^{n} \left\{ (B_t - C_t)/(1+r)^t \right\} \tag{4.5}$$

4.3.7.1. Internal Rate of Return (IRR)

Internal rate of return is the rate of return at which the Net Present value of a stream of payments/incomes is equal to zero.

$$IRR = \sum_{t=0}^{n} \left\{ (B_t - C_t)/(1+IRR)^t \right\} = 0 \tag{4.6}$$

4.3.7.2. Benefit Cost Ratio (BCR)

The benefit cost ratio (BCR) of an investment is the ratio of the discounted value of all cash inflows to the discounted value of all cash outflows during the life of the project. It can be estimated as follows:

$$BCR = \sum_{t=0}^{n} \left\{ (B_t)/(1+r)^t / \sum_{t=0}^{n} \left[(C_t)/(1+r)^t \right] \right\} \tag{4.7}$$

where,

B_t: Gross returns in time t

C_t: Variable cost in time t

r: Rate of interest

t: Time period (t = 0, 1, 2, ..i, ..30)

4.3.7.3. Pay-back Period

The Pay-back period is defined as the length of time required to recover an initial investment through cash flows generated by the investment.

$$\text{Pay Back Period} = \frac{\text{Cost of investment}}{\text{Annual net cash flow}}$$

4.3.8. Analysis of Marketing

The data collected were tabulated and analyzed for examining the marketing cost, margins, price spread and the marketing efficiency.

Marketing Margins, Costs and Loss

The post harvest loss at various stages of marketing has been included either in the farmer's net margin or market intermediaries margin. In the present study, the marketing loss at different stages has been explicitly estimated. The modified formulae has been used for separating the 'post harvest loss during marketing' at different stages of marketing as well as for estimating the producers' share, marketing margins and marketing loss.

(a) Net Farmers Price

The net price received by the grower has been estimated as the difference in gross price received and sum of marketing costs and value loss during harvesting, grading, transit and marketing. Thus, the net farmer's price is expressed mathematically as follows:

$$NP_F = GP_F - \{C_F + (L_F \times GP_F)\} \text{ or}$$

$$NP_F = \{GP_F\} - \{C_F\} - \{L_F \times GP_F\} \qquad (4.8)$$

where,

NP_F: Is net price received by the farmers (₹/kg),

GP_F: Is gross price received by the farmers or wholesale price to farmers (₹/kg),

C_F: Is the cost incurred by the farmers during marketing (₹/kg),

L_F: Is physical loss in produce from harvest till it reaches assembly market (per kg).

(b) Marketing Margins

The margins of market intermediaries included profit and returns, which accrued to them for storage, the interest on capital and establishment after adjusting for the marketing loss due to handling. The general expression for estimating the margin for intermediaries is given below.

Intermediaries margin = Gross price (sale price) − Price paid (cost price) − Cost of marketing − Loss in value during wholesaling

Net marketing margin of the wholesaler is given mathematically by

$$MM_W = GP_W - GP_F - C_W - (L_W \times GP_W) \text{ or}$$

$$MM_W = \{GP_W - GP_F\} - \{C_W\} - \{L_W \times GP_W\} \qquad (4.9)$$

where,

MM_W: Is net margin of the wholesaler (₹/kg),

GP_W: Is wholesaler's gross price to retailers or purchase price of retailer (₹/kg),

C_W: Is cost incurred by the wholesalers during marketing (₹/kg),

L_W: Is physical loss in the produce at the wholesale level (per kg).

In the marketing chain, when more than one wholesaler is involved, *i.e.*, primary wholesaler, secondary wholesaler, etc, then the total margin of the wholesaler is the sum of the margins of all wholesalers. Mathematically,

$$MM_w = MM_{w1} + ..+ MM_{wi} + ..+ MM_{wn}$$

where,

MM_{wi}: Is the marketing margin of the *i*-th wholesaler.

Net marketing margin of retailer is given by:

$$MM_R = GP_R - GP_W - C_R - (L_R \times GP_R) \text{ or}$$

$$MM_R = \{GP_R - GP_W\} - \{C_R\} - \{L_R \times GP_R\} \quad (4.10)$$

where,

MM_R: Is net margin of the retailer (₹/kg),

GP_R: Is price at the retail market or purchase price of the consumers (₹/kg),

L_R: Is physical loss in the produce at the retail level (per kg),

C_R: Is the cost incurred by the retailers during marketing (₹/kg).

The first bracketed term in equations (4.8), (4.9) and (4.10) indicates the gross return, while the second and third bracketed terms indicate respectively the cost and loss at different stages of marketing.

Thus, the total marketing margin of the market intermediaries (MM) is calculated as

$$MM = MM_W + MM_R \quad (4.11)$$

Similarly, the total marketing cost (MC) incurred by the producer/seller and by various intermediaries is calculated as

$$MC = C_F + C_W + C_R \quad (4.12)$$

Total loss in the value of produce due to injury/damage caused during handling of produce from the point of harvest till it reaches the consumers is estimated as

$$ML = \{L_F \times GP_F\} + \{L_W \times GP_W\} + \{L_R \times GP_R\} \quad (4.13)$$

(c) Marketing Efficiency

Most commonly used measures are conventional input to output marketing ratio, Shepherd's ratio of value (price) of goods marketed to the cost of marketing (Shephard, 1965) and Acharya's modified marketing efficiency formula (Acharya and Agarwal, 2001). However, all these measures do not explicitly mention the loss in the produce during the marketing process as a separates item in marketing. As reduction in loss itself is one of the efficiency parameters, there has been a need to incorporate this component explicitly in the existing marketing ratios to get correct measures of marketing efficiency while comparing alternate markets/channels. 'Marketing loss' component was incorporated in the widely used formula as given by Acharya and Agarwal (2001) and the modified marketing efficiency (ME) formula is given below.

$$ME = \frac{NP_F}{MM + MC + ML} \tag{4.14}$$

where,

NP_F: Is net price received by the farmers (₹/kg),

MM: Is the marketing margin,

MC: Is marketing cost,

ML: Is marketing loss.

4.3.9. Analysis of Seasonal Indices

$$S.I. = \frac{PI}{MA(12)} \times 100 \tag{4.15}$$

$$MA(12) = \frac{1}{12}\sum PI$$

where,

MA(12): Twelve month moving average

PI: Market arrivals/Price indices

S.I.: Seasonal Indices for market arrivals/prices

4.3.10. Production and Marketing Constraints

The selected orchardists were contacted through survey for analyzing the constraints faced by them at various levels. The information on different aspects of production and marketing constraints faced by the orchardists was tabulated into frequency tables and expressed in percentages against each of the item.

4.4. Results

The results pertaining to the case study, "*Economics of Production and Marketing of Citrus in Jammu region of Jammu and Kashmir State*" have been presented in this chapter through appropriate tables and figures.

4.4.1. Economics of Production of Orange

4.4.1.1. Resource Use Efficiency

For analysing the resource use efficiency of various factors, gross product had been taken as dependent variable with (manures + fertilizers), irrigation, plant protection, training/pruning and human labour as the independent variables. The age of the orchards (orange, kinnow and lemon) were grouped into six groups *viz.* 5th to 9th year, 10th to 14th year, 15th to 19th year, 20th to 24th year, 25th to 28th year and overall group which included the whole age of the orchard as was done by Wani *et al.* (1993). First to 4th year was not taken as these were the non bearing years. Various combinations of variables were tried. The choice of the best equation was made on the basis of R^2 explained and the relevance of the expected sign of coefficient. Some

variables like irrigation and plant protection (in some age groups) were omitted during the analysis as the sample orchardists had not used these resources in those age groups. The marginal value productivity (MVP) of these resources used was worked out with the help of regression coefficients obtained. MVP of a particular resource represents the expected addition to the gross return caused by the additional one unit of the resource input while the other inputs are held constant.

The regression function result and marginal value productivity of orange from 5th–9th year depicted in Table 4.3, indicated the output of orange orchards was regressed against human labour, manures + fertilizers, irrigation, plant protection and training/pruning. The crop production function for orange from 5th to 9th year was statistically significant having R^2 value (0.622) meaning that 62.2 per cent of the total variations in the production of orange was explained by the explanatory variables under consideration. The functional analysis for orange revealed that human labour was significant at 1 per cent level of probability with regression coefficient 0.771 and training/pruning was significant at 5 per cent level of probability with regression coefficient -0.017 whereas (manures + fertilizers), irrigation and plant protection with their values as 0.010, -0.014 and -0.010, respectively were non significant. The marginal value productivity of human labour (0.153) and manures + fertilizers (0.034) was positive. Irrigation, plant protection and training/pruning had negative marginal value productivity with their value at -603.966, -0.223 and -260.371, respectively.

Table 4.3: Estimated Regression Coefficients of Various Factors, their Standard Errors and MVP of Orange Production (5th–9th Year)

Variables	*Regression Coefficients*	*Standard Error*	*MVP*
Constant	-0.057	0.254	
Manures + Fertilizers	0.010	0.013	0.034
Irrigation	-0.014	0.007	-603.966
Plant Protection	-0.010	0.010	-0.223
Training/Pruning	-0.017**	0.008	-260.371
Human Labour	0.771*	0.080	0.153
F value	20.75		
Coefficient of determination (R^2)	**0.622***		

Note: * Significant at 1 per cent level of significance; **: Significant at 5 per cent level of significance.

Table 4.4 indicated regression function and marginal value productivity of orange for group II (10th–14th year). The R^2 (0.884), value was statistically significant meaning thereby that 88.4 per cent of the variation in the production of orange was due to the above mentioned explanatory variables. The data further indicated that the human labour and plant protection were found significant at 1 per cent and 5 per cent level of significance, respectively with regression coefficients as 0.867 and 0.045, whereas the variables (manures + fertilizers) with negative regression coefficient (-0.056) was statistically significant at 1 per cent level of probability. Training/pruning was found to be non significant. The marginal value productivity of human labour, plant

protection and training/pruning was positive with their corresponding values as 0.085, 1.354 and 512.188, respectively, whereas that of (manures + fertilizers) was negative (-1.845).

Table 4.4: Estimated Regression Coefficients of Various Factors, their Standard Errors and MVP of Orange Production (10th–14th year)

Variables	*Regression Coefficients*	*Standard Error*	*MVP*
Constant	-0.358**	0.145	
Manures + Fertilizers	-0.056*	0.014	-1.845
Plant Protection	0.045**	0.018	1.354
Training/Pruning	0.022	0.013	512.188
Human Labour	0.867*	0.040	0.085
F value	122.51		
Coefficient of determination (R^2)	**0.884****		

Note: * Significant at 1 per cent level of significance; **: Significant at 5 per cent level of significance.

The regression function result and marginal value productivity of orange from 15th–19th year presented in Table 4.5 depicted statistically significant R^2 (0.890), thereby meaning that 89.0 per cent of the total variation in the production of orange was explained by the above mentioned explanatory variables like (manures + fertilizers), plant protection, training/pruning and human labour. Only one variable *i.e.*, human labour was statistically significant at 1 per cent level of probability, with regression coefficient of 0.874, whereas the rest of the variables were non significant having regression coefficient of 0.001 (manures + fertilizers), -0.002 (plant protection) and –0.015 (training/pruning). The marginal value productivity of human labour (0.106) and manures + fertilizers (0.040) was positive whereas plant protection (-0.995) and training/pruning (-622.236) was negative.

Table 4.5: Estimated Regression Coefficients of Various Factors, their Standard Errors and MVP of Orange Production (15th–19th Year)

Variables	*Regression Coefficients*	*Standard Error*	*MVP*
Constant	-0.532*	0.158	
Manures + Fertilizers	0.001	0.014	0.040
Plant Protection	-0.002	0.012	-0.995
Training/Pruning	-0.015	0.013	-622.236
Human Labour	0.874*	0.040	0.106
F value	129.03		
Coefficient of determination (R^2)	**0.890***		

Note: * Significant at 1 per cent level of significance; **: Significant at 5 per cent level of significance.

The regression function result and marginal value productivity of orange from 20th–24th year presented in Table 4.6 depicted that the output of orange orchards was regressed against human labour, (manures + fertilizers) and plant protection. The crop production function used was statistically significant having R^2 value as 0.733 meaning that 73.3 per cent of the total variation in the production of orange was explained by the explanatory variables under consideration. The regression coefficient for human labour was positive and significant at 1 per cent level of probability (0.889). Rest of the two variables (manures + fertilizers) and plant protection with regression coefficients as 0.054 and -0.052, respectively were non significant. The marginal value productivity of the human labour (0.098) and (manures + fertilizers) as 59.545 was positive whereas that of plant protection was negative with its value at -0.003.

Table 4.6: Estimated Regression Coefficients of Various Factors, their Standard Errors and MVP of Orange Production (20th–24th year)

Variables	*Regression Coefficients*	*Standard Error*	*MVP*
Constant	-1.178**	0.567	
Manures + Fertilizers	0.054	0.057	59.545
Plant Protection	-0.052	0.103	-0.003
Human Labour	0.889*	0.074	0.098
F value	43.86		
Coefficient of determination (R^2)	**0.733***		

Note: * Significant at 1 per cent level of significance; **: Significant at 5 per cent level of significance.

The regression function result and marginal value productivity of orange from 25th–28th year presented in Table 4.7 depicted that the output of orange orchards was regressed against human labour, (manures + fertilizers) and plant protection. The data revealed that crop production function used was statistically significant having R^2 value (0.678) meaning thereby that 67.8 per cent of the total variation in the production of orange was explained by the explanatory variables under consideration. The functional analysis for orange production revealed that human labour and (manures + fertilizers) were positively significant at 1 per cent level of probability with regression coefficients as 0.723 and 0.173, respectively. The table further revealed that plant protection was negatively non significant with its regression coefficient as -0.158. The marginal value productivity of human labour and (manures + fertilizers) was however positive with their values at 0.365 and 203.462, respectively, whereas that of plant protection was negative with its value at -0.010.

The regression function result and marginal value productivity of overall orange orchards *i.e.* from its pre bearing period to maturity period (1st year–28th year) presented in Table 4.8 depicted that the output of orange orchards was regressed against human labour, (manures + fertilizers), irrigation, plant protection and training/pruning. The production function used was statistically significant having R^2 value (0.687), which indicated that 68.7 per cent of the total variations in the production function for orange was explained by the above mentioned explanatory variables. The variables

like human labour and training/pruning were found to be significant at 1 per cent level of probability with regression coefficients as 0.955 and -0.050, respectively. The other three explanatory variables *i.e.* (manures + fertilizers), irrigation and plant protection were non significant with their regression values at 0.012, -0.012 and 0.013, respectively. The marginal value productivity of manures + fertilizers, human labour and plant protection was positive with their values at 110.452, 0.185 and 0.076, respectively, whereas that of irrigation, training/pruning was negative with their values at -0.054 and -0.638, respectively.

Table 4.7: Estimated Regression Coefficients of Various Factors, their Standard Errors and MVP of Orange Production (25th–28th Year)

Variables	*Regression Coefficients*	*Standard Error*	*MVP*
Constant	-0.582	0.635	
Manures + Fertilizers	0.173*	0.064	203.462
Plant Protection	-0.158	0.115	-0.010
Human Labour	0.723*	0.064	0.365
F value	33.76		
Coefficient of determination (R^2)	**0.678***		

Note: * Significant at 1 per cent level of significance; **: Significant at 5 per cent level of significance.

Table 4.8: Estimated Regression Coefficients of Various Factors, their Standard Errors and MVP of Orange Production (Overall)

Variables	*Regression Coefficients*	*Standard Error*	*MVP*
Constant	-0.399	0.387	
Manures + Fertilizers	0.012	0.013	110.452
Irrigation	- 0.012	0.009	-0.054
Plant Protection	0.013	0.011	0.076
Training/Pruning	-0.050*	-0.017	-0.638
Human Labour	0.955*	0.094	0.185
F value	27.61		
Coefficient of determination (R^2)	**0.687***		

Note: * Significant at 1 per cent level of significance; **: Significant at 5 per cent level of significance.

4.4.1.2. Costs and Returns

Orange is a perennial fruit crop and the plants start bearing five years after planting in Jammu region. During the productive life of the tree, it continues to bear fruits and yields sizeable income to the growers. Orange requires a relatively high investment of capital in the first year for the establishment of the orchard followed by relatively low costs in the subsequent years.

(a)

(b)

(c)

Figure 4.2 (a-c): Glimpses of Citrus Orchards in Study Area
(a) Orchard of Sh. Ram Dev of Village Nai Basti Block Marh;
(b) Orchard of Sh. Shiv Das at Village Amli Block Samba;
(c) Kinnow orchard of Sh. Jyoti Sawroop at Village Nai Basti Block Marh

The information with respect to yearly costs involved in the establishment of orange orchards was necessary in order to allocate the non recurring cost to the yearly cost of production.

4.4.1.2.1 Establishment Cost

During the first year of the plantation, orchardists had incurred expenditure on the preparation of land, planting material, digging, filling of pits and planting, irrigation, plant protection, (manures + fertilizers) etc. The operation wise per acre costs for different size groups of holdings for the first year is presented in Table 4.9 and Figure 4.3. The per acre total costs in first year for the establishment of orange orchards were ₹5018.76 for marginal orchards, ₹5087.62 for small orchards, ₹5362.70 for medium orchards, ₹5590.11 for large orchards. The Marginal orchardist incurred ₹1395.85 on digging, filling and planting, ₹922.52 on preparation of land, ₹407.83 on planting material, ₹57.35 on irrigation, ₹35.00 on training/pruning, ₹7.59 on plant protection and ₹203.24 on (manures + fertilizers) with depreciation of ₹123.50 and earned value of rented land (EVRL) of ₹1310.47 per acre. The per acre interest on working and fixed capital was worked out to ₹383.33 and ₹172.08, respectively. The per acre costs incurred by small orchardists were ₹1422.68 on digging, filling and planting, ₹995.50 on preparation of land, ₹373.78 on planting material, ₹56.98 on irrigation, ₹40.00 on training/pruning ₹6.95 on plant protection and ₹170.03 on (manures + fertilizers) with depreciation of ₹142.25 and EVRL of ₹1349.35. The interest on working and fixed capital was worked out to ₹351.11 and ₹178.99, respectively. Similarly, per acre costs incurred in medium orchards were ₹1385.94 on digging, filling and planting, ₹1010.11 on preparation of land, ₹515.78 on planting material, ₹55.35 on irrigation, ₹48.00 on training/pruning, ₹5.35 on plant protection and ₹158.78 on (manures + fertilizers) with depreciation of ₹165.21 and EVRL of ₹1427.32. The interest on working and fixed capital was worked out to ₹399.76 and ₹191.10, respectively while as in case of large orchards, the per acre costs incurred were ₹1463.84 on digging, filling and planting, ₹1213.31 on preparation of land, ₹490.52 on planting material, ₹58.68 on irrigation, ₹48.00 on training/pruning, ₹6.15 on plant protection and ₹164.77 on manures + fertilizers with depreciation of ₹125.36 and EVRL of ₹1464.97. The interest on working and fixed capital was worked out to ₹363.67 and ₹190.84, respectively. The average per acre first year establishment costs were ₹5089.08 out of which ₹1405.79 were incurred on digging, filling and planting, ₹962.21 on preparation of land, ₹411.61 on planting material, ₹57.28 on irrigation, ₹37.52 on training/pruning, ₹7.26 on plant protection and ₹191.78 on manures + fertilizers, ₹129.25 as depreciation on the machinery and farm inventory and EVRL of ₹1334.63. Similarly, the per acre interest on working and fixed capital was worked out to ₹376.08 and ₹175.67, respectively.

The year wise per acre establishment cost of orange orchards are presented in Table 4.10 and Figure 4.4. The total establishment cost per acre incurred on orange were ₹12077.03 in marginal orchards, ₹12542.27 in small orchards, ₹13198.00 in medium orchards and ₹13691.59 in large orchards. The year wise establishment cost in marginal orchards were ₹5018.76 in the first year, ₹2267.08 in the second year, ₹2367.10 in the third year and ₹2424.09 in the fourth year. In small orchards the year wise per acre establishment costs were ₹5087.62 during the first year, with subsequent

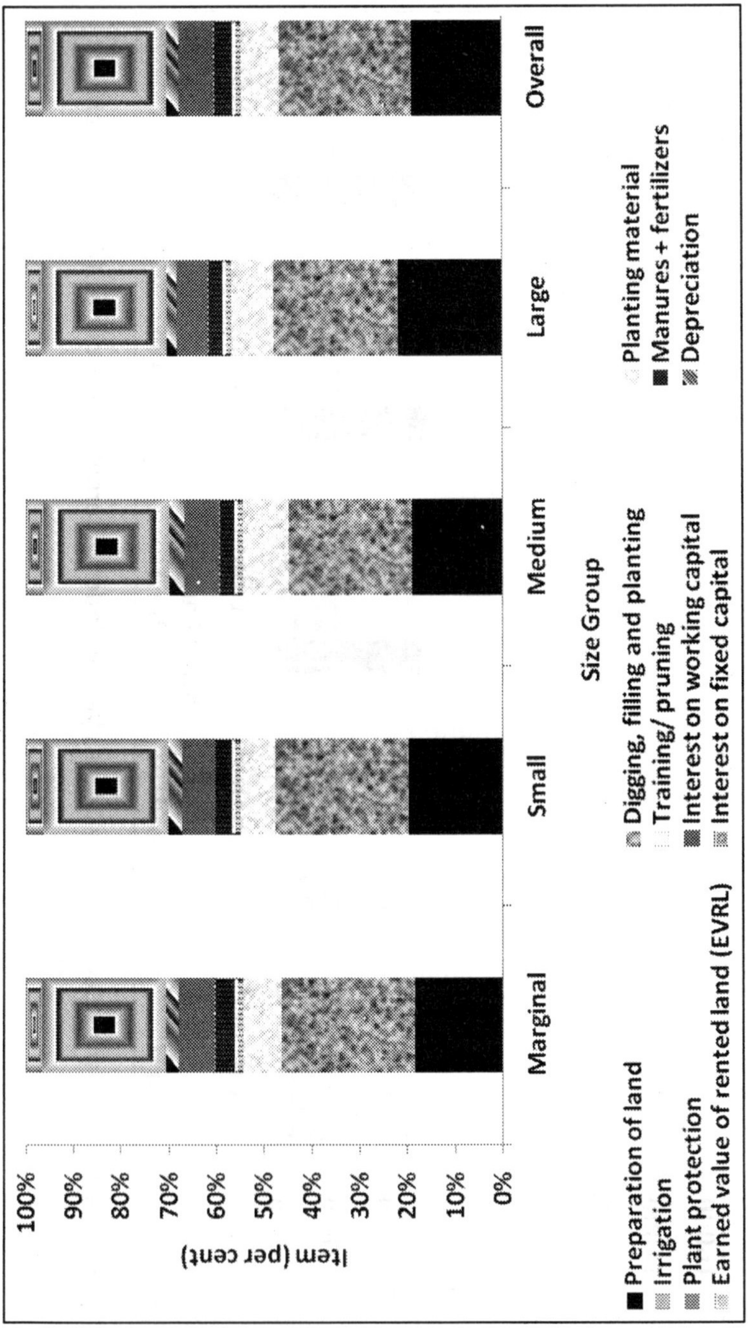

Figure 4.3: Per acre First Year Establishment Costs (Per cent) Under Different Size Groups of Orange Orchards

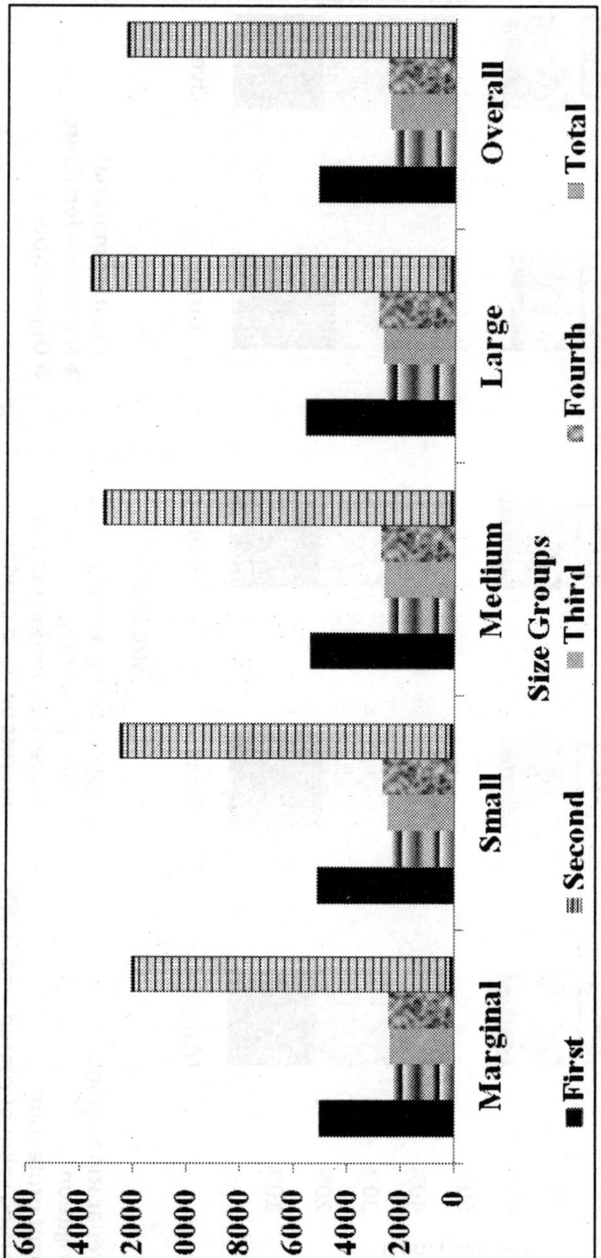

Figure 4.4: Year-wise Establishment Costs (₹/acre) Under Different Size Groups of Orange Orchards

costs for second, third and fourth year as ₹2337.36, ₹2477.74 and ₹2639.55, respectively. The data further indicated that per acre year wise establishment costs incurred by medium orchardists were ₹5362.70 in the first year, ₹2501.20 in the second year, ₹2598.00 in the third year and ₹2736.10 in the fourth year. Similarly, for the large orchardists, it was ₹5590.11 in the first year, ₹2589.73 in the second year, ₹2669.92 in the third year and ₹2841.83 in the fourth year. The perusal of the table further revealed that on an average, the total per acre establishment costs for all categories together were ₹12337.16 out of which ₹5089.00, ₹2314.90, ₹2421.53 and ₹2511.64 were realized during I, II, III and IV year of establishment costs, respectively.

Table 4.9: Operation-wise First Year Establishment Costs Under Different Size Groups of Orange Orchards

(₹/acre)

Item	*Marginal*	*Small*	*Medium*	*Large*	*Overall*
Preparation of land	922.52	995.50	1010.11	1213.31	962.21
Digging, filling and planting	1395.85	1422.68	1385.94	1463.84	1405.79
Planting material	407.83	373.78	515.78	490.52	411.61
Irrigation	57.35	56.98	55.35	58.68	57.28
Training/pruning	35.00	40.00	48.00	48.00	37.52
Manures + fertilizers	203.24	170.03	158.78	164.77	191.78
Plant protection	7.59	6.95	5.35	6.15	7.26
Interest on working capital	383.33	351.11	399.76	363.67	376.08
Land revenue	0.00	0.00	0.00	0.00	0.00
Depreciation	123.50	142.25	165.21	125.36	129.25
Earned value of rented land (EVRL)	1310.47	1349.35	1427.32	1464.97	1334.63
Interest on fixed capital	172.08	178.99	191.10	190.84	175.67
Total	**5018.76**	**5087.62**	**5362.70**	**5590.11**	**5089.08**

Table 4.10: Year-wise Establishment Costs Under Different Size Groups of Orange Orchards

(₹/acre)

Year	*Marginal*	*Small*	*Medium*	*Large*	*Overall*
I	5018.76	5087.62	5362.70	5590.11	5089.09
II	2267.08	2337.36	2501.20	2589.73	2314.90
III	2367.10	2477.74	2598.00	2669.92	2421.53
IV	2424.09	2639.55	2736.10	2841.83	2511.64
Total	**12077.03**	**12542.27**	**13198.00**	**13691.59**	**12337.16**

4.4.1.2.2. Operational Costs

The per acre operational costs consists of yearly expenses on the maintenance of bearing orchard. The item wise and concept wise operational costs for orange orchard are presented in Table 4.11 and Figures 4.5 and 4.6. The per acre cost A which included all the variable costs excluding the family human labour were ₹1008.91 for marginal orchards, ₹1110.42 for small orchards, ₹1356.15 for medium orchards and ₹1233.58 for large orchards, whereas on an average it was ₹1060.88. The per acre cost B which included the fixed costs in addition to cost A were ₹2614.96 for marginal orchards, ₹2781.01 for small orchards, ₹3139.78 for medium orchards and ₹3014.75 for large orchards with an average of ₹2700.44. The cost C *i.e.* total item wise per acre operational costs which also included the imputed value of family labour were ₹3714.96 for marginal orchards, ₹4031.01 for small orchards, ₹4439.78 for medium orchards and ₹4249.75 for large orchards with an average of ₹3849.35.

Table 4.11: Item-wise and Concept-wise Operational Costs Under Different Size Groups of Orange Orchards

(₹/acre)

Sl.No.	*Item*	*Marginal*	*Small*	*Medium*	*Large*	*Overall*
1.	Hired human labour	555.34	565.25	691.75	552.70	563.09
2.	Irrigation	0.00	0.00	0.00	0.00	0.00
3.	Training/pruning	35.69	49.32	58.80	74.08	42.24
4.	Manures + fertilizers	295.24	352.99	423.44	438.13	322.89
5.	Plant protection	14.54	23.89	36.86	36.50	19.00
6.	Interest on working capital	108.10	118.97	145.30	132.17	113.67
7.	Land revenue	0.00	0.00	0.00	0.00	0.00
8.	Depreciation	123.50	142.25	165.21	125.36	129.25
9.	EVRL*	1310.47	1349.35	1427.32	1464.97	1334.63
10.	Interest on fixed capital	172.08	178.99	191.10	190.84	175.67
11.	Family human labour	1100.00	1250.00	1300.00	1235.00	1148.91
12.	Cost A (1-6)	**1008.91**	**1110.42**	**1356.15**	**1233.58**	**1060.88**
13.	Cost B (1-10)	**2614.96**	**2781.01**	**3139.78**	**3014.75**	**2700.14**
14.	Cost C (1-11)	**3714.96**	**4031.01**	**4439.78**	**4249.75**	**3849.35**

*: Earned value of rented land.

The Table 4.11 further revealed that among the working costs, the various expenditures incurred by different size of holding were ₹555.34 on hired human labour, ₹35.69 on training/pruning, ₹295.24 on (manures + fertilizers), ₹14.54 on plant protection and ₹108.10 as interest on working capital in case of marginal orchards, whereas it was ₹565.26, ₹49.32, ₹352.99, ₹23.89 and ₹118.97, respectively in case of small orchards, ₹691.75, ₹58.80, ₹423.44, ₹36.86 and ₹145.30, respectively in case of medium orchards and ₹552.70, ₹74.08, ₹438.13, ₹36.50 and ₹132.17, respectively in case of large orchards. The fixed costs included ₹123.50 on depreciation on machinery and farm inventory, ₹1310.47 as EVRL (Earned value of

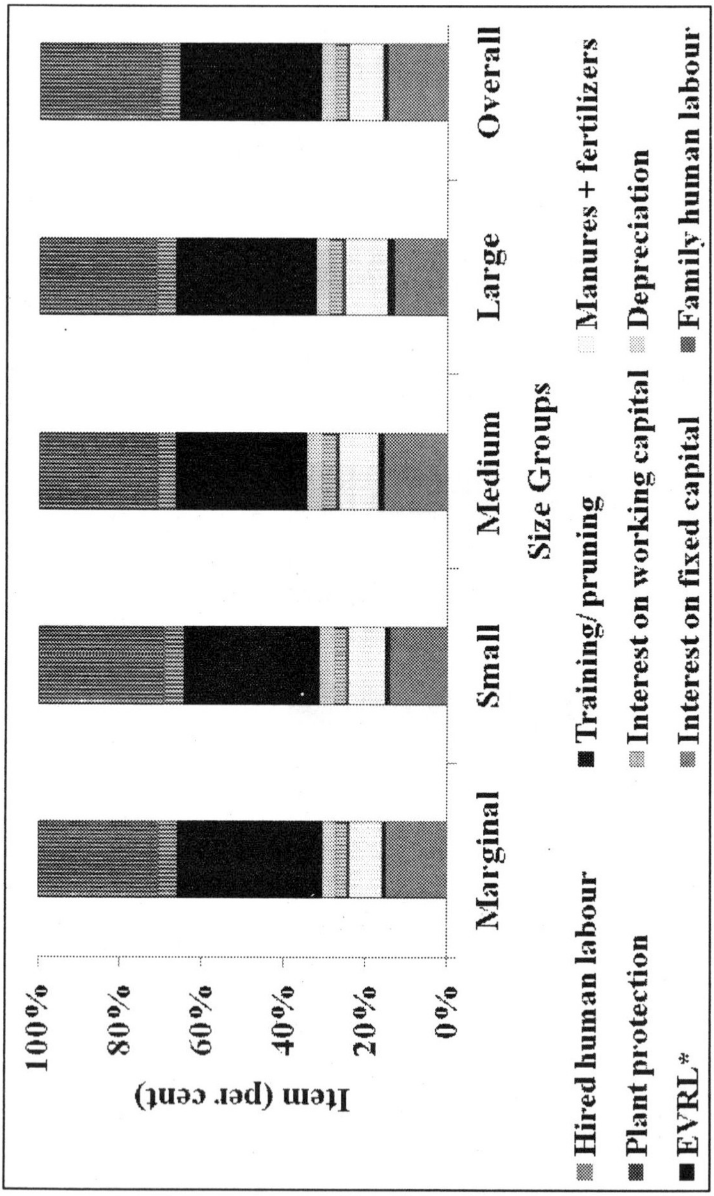

Figure 4.5: Item-wise per Acre Operational Costs (Per cent) Under Different Size Groups of Orange Orchards

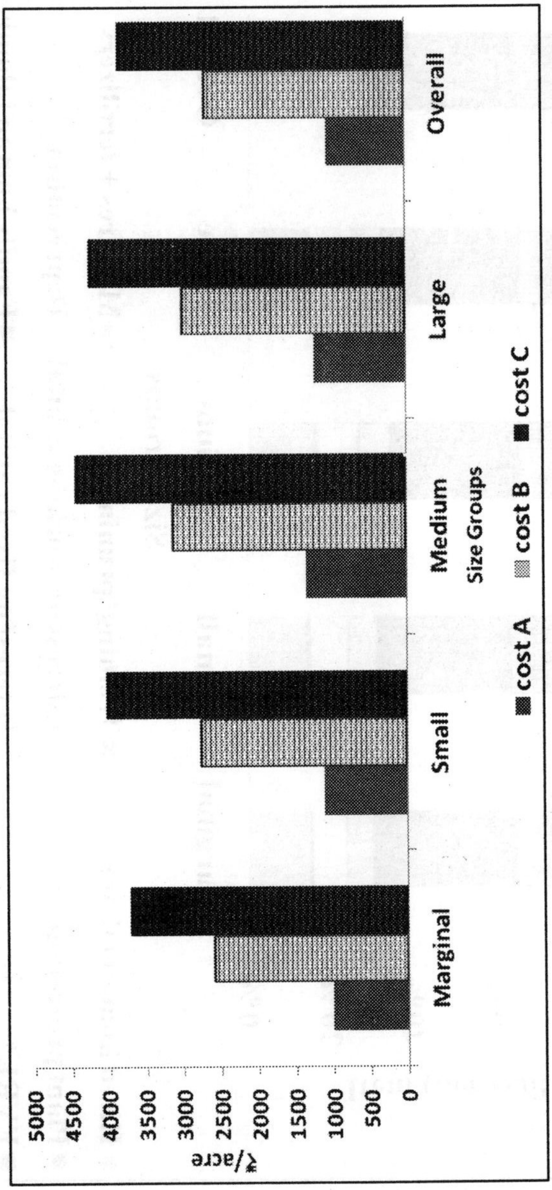

Figure 4.6: Concept-wise Cost of Production (₹/acre) Under Different Size Groups of Orange Orchards

rented land) and ₹172.08 as interest on fixed capital in case of marginal orchards, ₹142.25, ₹1349.35 and ₹178.99, respectively in case of small orchards, ₹165.21, ₹1427.32 and ₹191.10, respectively in case of medium orchards and ₹125.36, ₹1464.97 and ₹190.84, respectively in case of large orchards. The per acre family human labour used was of the value of ₹1100.00, ₹1250.00, ₹1300.00 and ₹1235.00, respectively in marginal, small, medium and large orchards. On an average, the per acre expenditures were ₹563.09 on hired human labour, ₹1148.91 as family human labour, ₹322.89 as (manures + fertilizers), ₹1334.63 as earned value of rented land, ₹19.00 as plant protection and ₹42.24 as training/pruning.

4.4.1.2.3. Returns

The returns per year from orange orchards per acre worked out for different age groups are presented in Table 4.12 and Figure 4.7. The returns were worked out upto 10th year of the age of an orchard, between 11th – 15th year and above 15th year of orchard. These groups were made on the fact that the returns upto 10th year were low as compared to other groups, whereas from 11th – 15th year the returns were increasing and from 16th year onwards the returns almost remained constant.

The returns upto10th year of age were ₹6963.67 in marginal orchards, ₹7142.32 in small orchards, ₹7287.07 in medium orchards and ₹7325.65 in case of large orchards per acre per year. The returns per acre per year from 11th – 15th year were ₹7895.35 in marginal orchards, ₹8563.25 in small orchards, ₹8965.34 in medium orchards and ₹9152.21 in case of large orchards, while as from 15th year onwards, the returns per acre per year were ₹8265.35 in marginal orchards, ₹8356.59 in small orchards, ₹8998.86 in medium orchards and ₹9326.21 in large orchards. The average returns per acre per year from marginal, small, medium and large orchards were ₹7862.85, ₹8096.08, ₹8563.93 and ₹8789.82, respectively.

4.4.1.3. Economic Viability

The economic viability of orange orchards is presented in Table 4.13. The net present value was worked out to ₹3347.62 in marginal orchards, ₹5984.83 in small orchards, ₹5330.86 in medium orchards and ₹4130.40 in large orchards. The internal rate of return was 12.56 per cent in marginal orchards, 16.50 per cent in small orchards, 15.33 per cent in medium orchards and 18.25 per cent in case of large orchards. The benefit cost ratio was 2.01, 2.14, 2.33 and 2.33 in marginal, small, medium and large orchards, respectively with pay-back period of 7.2, 7.5, 8.1, and 8.1 respectively. On an average the net present value of orange orchards was ₹4025.66 having internal rate of return 15.66 per cent, benefit cost ratio 2.07 and having a pay-back period of 7.4 years.

4.4.2. Economics of Production of Kinnow

4.4.2.1. Resource Use Efficiency

In order to estimate the resource productivities and thereby to establish the functional relationship between the yields of kinnow as the dependent variable and the selected independent variables, log linear function was fitted to the data collected for the production of kinnow for different ages of the orchards separately. The age of the orchards were grouped on the similar pattern as was done for orange orchards.

Various combinations of variables were tried. The choice of the best equation was made on the basis of R^2 explained. Some variables like irrigation and training/pruning in some age groups were omitted during the analysis as the sample orchardists had not used these resources in those age groups.

Table 4.12: Average Returns Under Different Age Groups of Orange Orchards

(₹/acre/year)

Item	*Marginal*	*Small*	*Medium*	*Large*	*Overall*
Upto 10 years	6963.67	7142.32	7287.07	7325.65	7040.21
11th to 15th year	7895.35	8563.25	8965.34	9152.21	8168.46
Above 15th year	8265.35	8356.59	8998.86	9326.21	8392.63
Overall	**7862.85**	**8096.08**	**8563.93**	**8789.82**	**8007.82**

Table 4.13: Economic Viability Under Different Size Groups of Orange Orchards

Group	*Net Present Value (₹)*	*Internal Rate of Return (per cent)*	*Payback Period (Years)*	*Benefit Cost Ratio*
Marginal	3347.62	12.56	7.2	2.01
Small	5984.83	16.50	7.5	2.14
Medium	5330.86	15.33	8.1	2.33
Large	4130.40	18.25	8.1	2.33
Overall	**4025.66**	**15.66**	**7.4**	**2.07**

Table 4.14: Estimated Regression Coefficients of Various Factors, their Standard Errors and MVP of Kinnow Production (5th–9th Year)

Variables	*Regression Coefficients*	*Standard Error*	*MVP*
Constant	0.796*	0.262	
Manures + Fertilizers	0.320*	0.073	0.469
Irrigation	0.004	0.024	2.047
Plant Protection	-0.006	0.038	-0.104
Training/Pruning	0.030	0.030	0.064
Human Labour	0.316*	0.094	0.097
F value	19.37		
Coefficient of determination (R^2)	**0.787***		

Note: * Significant at 1 per cent level of significance; **: Significant at 5 per cent level of significance.

The regression function result and marginal value productivity of kinnow from 5th–9th year presented in Table 4.14 indicated that the output of kinnow orchards was regressed against human labour, (manures + fertilizers), irrigation, plant protection

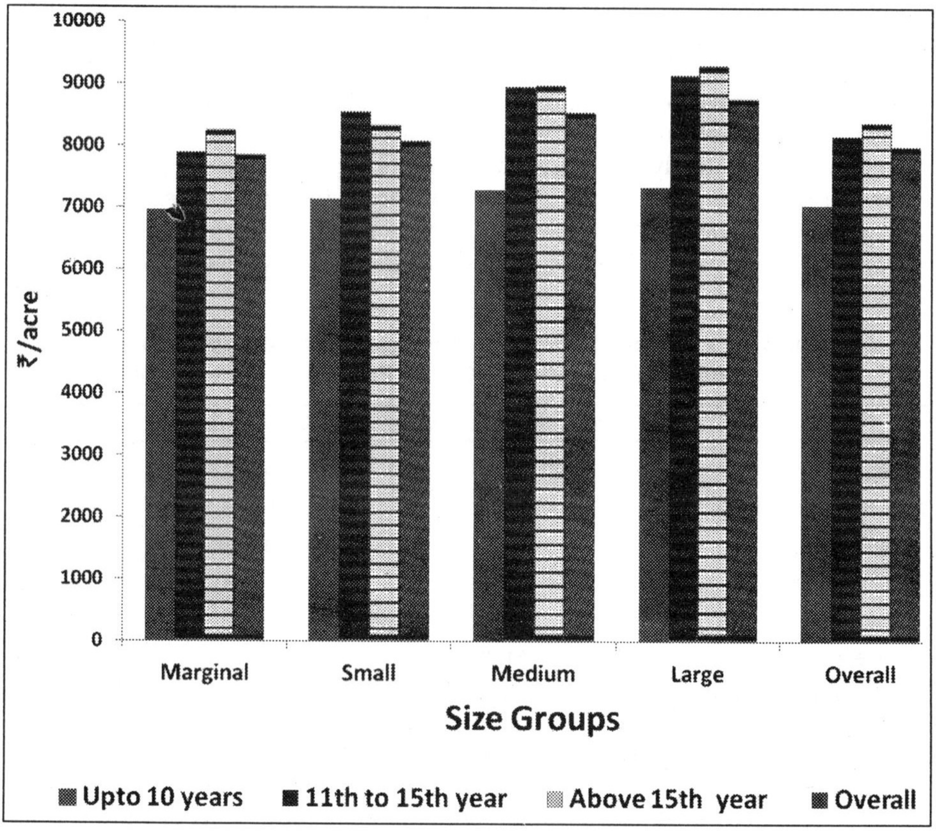

Figure 4.7: Average Returns Under Different Age Groups (₹/acre) of Orange Orchards

and training/pruning. The crop production function used from 5th – 9th year of establishment was found statistically significant having R^2 value (0.787) meaning that 78.7 per cent of the total variations in the production function for kinnow was explained by the explanatory variables under consideration. The functional analysis for kinnow production revealed that human labour and (manures + fertilizers) were positively significant at 1 per cent level of probability with regression coefficients as 0.316 and 0.320, respectively whereas irrigation, plant protection and training/pruning were non significant with regression coefficients as 0.004, -0.006 and 0.030, respectively. Whereas the marginal value productivity of human labour (0.097), (manures + fertilizers as 0.469), irrigation (2.047) and training/pruning (0.064) was positive and that of plant protection (-0.104) was negative.

Table 4.15 presented the data on regression function and marginal value productivity of kinnow from 10th–14th year and revealed that kinnow orchards were regressed against human labour, (manures + fertilizers), plant protection and training/pruning. Under this age group, the R^2 value worked out to be 0.760, thereby indicated that 76.0 per cent of the total variation in the production of kinnow from 10th – 14th

year was explained by the above mentioned explanatory variables. The independent variables *viz.* human labour and (manures + fertilizers) were found to be positively significant at 1 per cent level of probability with regression coefficients as 0.245 and 0.573, respectively, whereas training/pruning was found to be positively significant at 5 per cent level of probability with regression coefficient as 0.124. The plant protection was negatively non significant with its regression coefficient as –0.005. The marginal value productivity of human labour, (manures + fertilizers) and training/pruning was positive with their corresponding values as 0.003, 0.996 and 0.009, respectively, whereas that of plant protection was negative with its value at –0.0002.

Table 4.15: Estimated Regression Coefficients of Various Factors, their Standard Errors and MVP of Kinnow Production (10th–14th Year)

Variables	*Regression Coefficients*	*Standard Error*	*MVP*
Constant	0.884*	0.382	
Manures + Fertilizers	0.573*	0.065	0.996
Plant Protection	-0.005	0.063	-0.0002
Training/Pruning	0.124**	0.059	0.009
Human Labour	0.245*	0.062	0.003
F Value	81.33		
Coefficient of determination (R^2)	**0.760***		

Note: * Significant at 1 per cent level of significance; **: Significant at 5 per cent level of significance.

The regression function result and marginal value productivity of kinnow from 15th – 19th year presented in Table 4.16 depicted that the output of kinnow orchards was regressed against human labour, (manures + fertilizers), plant protection and training/pruning and the production function used was statistically significant having R^2 value as 0.889 meaning that 88.9 per cent of the total variations in the production function for kinnow was explained by the above mentioned explanatory variables. The functional analysis for kinnow production revealed that human labour, manures + fertilizers and plant protection were found to be significant at 5 per cent level of probability with regression coefficient as 0.226, 0.263 and 0.137, respectively and training/pruning was positive but non significant with regression coefficient as 0.026. The marginal value productivity of human labour, manures + fertilizers, plant protection and training/pruning was positive with their values at 0.012, 9.071, 0.021 and 0.007, respectively.

The regression function result and marginal value productivity of kinnow from 20th – 24th year presented in Table 4.17 depicted that the output of kinnow orchards was related against human labour, manures + fertilizers and plant protection and the production function used was statistically significant having R^2 value as 0.872 meaning that 87.2 per cent of the total variation in the production for kinnow was explained by the above mentioned explanatory variables. The functional analysis for kinnow production revealed that only manures + fertilizers was found to be significant

at 1 per cent level of probability with regression coefficients as 0.710 whereas, other variables *viz.* human labour and plant protection were positive but non significant with regression coefficients as 0.005 and 0.156, respectively. The marginal value productivity of human labour, manures + fertilizers and plant protection was positive with their values at 0.020, 0.004 and 0.006, respectively.

Table 4.16: Estimated Regression Coefficients of Various Factors, their Standard Errors and MVP of Kinnow Production (15th–19th Year)

Variables	*Regression Coefficients*	*Standard Error*	*MVP*
Constant	1.837*	0.288	
Manures + Fertilizers	0.263**	0.108	9.071
Plant Protection	0.137**	0.064	0.021
Training/Pruning	0.026	0.069	0.007
Human Labour	0.226**	0.097	0.012
F value	24.65		
Coefficient of determination (R^2)	**0.889****		

Note: * Significant at 1 per cent level of significance; **: Significant at 5 per cent level of significance.

Table 4.17: Estimated Regression Coefficients of Various Factors, their Standard Errors and MVP of Kinnow Production (20th–24th Year)

Variables	*Regression Coefficients*	*Standard Error*	*MVP*
Constant	1.619**	0.643	
Manures + Fertilizers	0.710*	0.105	0.004
Plant Protection	0.156	0.135	0.006
Human Labour	0.005	0.090	0.020
F value	46.29		
Coefficient of determination (R^2)	**0.872***		

Note: * Significant at 1 per cent level of significance; **: Significant at 5 per cent level of significance.

The regression function result and marginal value productivity of kinnow from 25th–28th year presented in Table 4.18 depicted that the output of kinnow orchards was regressed against human labour, manures + fertilizers and plant protection and the production function used was statistically significant having R^2 value (0.727) meaning that 72.7 per cent of the total variations in the production for kinnow was explained by the explanatory variables under consideration. Moreover, the human labour was found to be significant at 1 per cent level of probability with regression coefficients as 0.751, whereas manures + fertilizers and plant protection were non significant with regression coefficients 0.025 and -0.143, respectively. The marginal value productivity of human labour and manures + fertilizers was positive with their

values at 0.498 and 0.042, respectively, whereas that of plant protection was negative with its value at -1.079.

Table 4.18: Estimated Regression Coefficients of Various Factors, their Standard Errors and MVP of Kinnow Production (25th–28th Year)

Variables	*Regression Coefficients*	*Standard Error*	*MVP*
Constant	-0.575	0.785	
Manures + Fertilizers	0.025	0.084	0.042
Plant Protection	-0.143	0.162	-1.079
Human Labour	0.751*	0.086	0.498
F value	58.35		
Coefficient of determination (R^2)	**0.727***		

Note: * Significant at 1 per cent level of significance; **: Significant at 5 per cent level of significance.

Table 4.19: Estimated Regression Coefficients of Various Factors, their Standard Errors and MVP of Kinnow Production (Overall)

Variables	*Regression Coefficients*	*Standard Error*	*MVP*
Constant	2.991*	0.084	
Manures + Fertilizers	-0.024	0.021	-0.027
Irrigation	0.016	0.011	0.025
Plant Protection	0.015	0.019	0.014
Training/Pruning	0.138*	0.011	0.175
Human Labour	0.029**	0.014	0.031
F value	35.61		
Coefficient of determination (R^2)	**0.736***		

Note: * Significant at 1 per cent level of significance; **: Significant at 5 per cent level of significance.

The regression function result and marginal value productivity of overall kinnow orchards presented in Table 4.19 depicted that the output of kinnow orchards was regressed against human labour, manures + fertilizers, irrigation, plant protection and training/pruning. The production function used was statistically significant having R^2 value as 0.736 meaning that 73.6 per cent of the total variations in the production function for kinnow was explained by the explanatory variables under consideration. The functional analysis for kinnow production revealed that human labour was found to be significant at 5 per cent level of probability with regression coefficient as 0.029 and training/pruning was found to be positively significant at 1 per cent level of probability with regression coefficient value at 0.138. The variables like irrigation and plant protection were found to be positive and non significant with regression coefficient as 0.016 and 0.015, respectively and manures + fertilizers was positive but non significant with regression coefficient as -0.024. The marginal

value productivity of human labour, irrigation, plant protection and training/pruning was positive with their values at 0.031, 0.025, 0.014 and 0.175, respectively whereas that of manures + fertilizers was negative with its value at -0.027.

4.4.2.2. Costs and Returns

Kinnow being a perennial fruit crop starts bearing five years after plantation in the Jammu region of Jammu and Kashmir state. The information with respect to yearly costs involved in the establishment of kinnow orchards is necessary in order to allocate the non recurring cost to the yearly cost of production.

4.4.2.2.1 Establishment Cost

Like orange, high investment is involved for the establishment of kinnow orchard during its first year with low cost involvement during the subsequent years till it starts bearing. The operation wise first year establishment costs of kinnow is presented in Table 4.20 and Figure 4.8. In case of marginal orchards, the per acre costs incurred were ₹1105.62 on preparation of land, ₹1630.00 on digging, filling and planting, ₹341.71 on planting material, ₹95.88 on irrigation, ₹38.00 on training/pruning, ₹254.30 on (manures + fertilizers), ₹125.36 as depreciation on machinery and farm inventory and ₹1093.69 on earned value of rented land (EVRL). The interest on working and fixed capital worked out to be ₹431.65 and ₹146.29, respectively, hence the total costs amounted to ₹5262.50. In case of small orchards, ₹1601.19 were spent on digging, filling and planting, ₹1154.18 on preparation of land, ₹379.69 on planting material, ₹84.38 on irrigation, ₹38.00 on training/pruning, ₹11.40 on plant protection and ₹259.13 on (manures + fertilizers). The depreciation on machinery and farm inventory

Table 4.20: Operation-wise First Year Establishment Costs Under Different Size Groups of Kinnow Orchards

(₹/acre)

Item	*Marginal*	*Small*	*Medium*	*Overall*
Preparation of land	1105.62	1154.18	1010.65	1113.77
Digging, filling and planting	1630.00	1601.19	1525.47	1620.69
Planting material	341.71	379.69	358.03	350.60
Irrigation	95.88	84.38	105.15	93.58
Training/Pruning	38.00	38.00	40.00	38.06
Manures + Fertilizers	254.3	259.13	432.88	260.33
Plant protection	0.00	11.4	46.59	3.83
Interest on working capital	431.65	456.8	441.45	437.51
Land revenue	0.00	0.00	0.00	0.00
Depreciation	125.36	126.58	165.26	126.74
Earned value of rented land (EVRL)	1093.69	1129.00	1230.9	1105.35
Interest on fixed capital	146.29	150.65	167.54	147.85
Total	5262.50	5391.00	5523.92	5298.32

Note: There was not a single large orchardist in case of kinnow fruit cultivation.

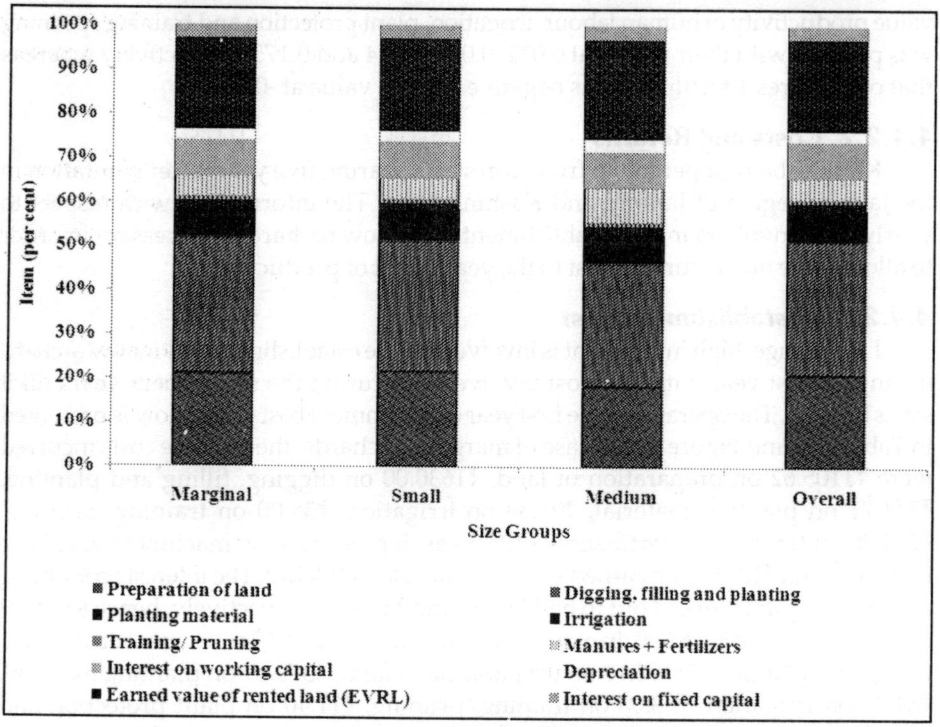

Figure 4.8: Per Acre First Year Establishment Costs (Per cent) Under Different Size Groups of Kinnow Orchards

was ₹126.58 and EVRL worked out to be ₹1129.00 with interest on working and fixed capital to be ₹456.80 and ₹150.65, respectively. The costs incurred by medium orchardists were ₹1525.47 on digging, filling and planting, ₹1010.65 on preparation of land, ₹358.03 on planting material, ₹105.15 on irrigation, ₹40.00 on training/ pruning, ₹46.59 on plant protection and ₹432.88 on (manures + fertilizers) with depreciation of ₹165.26 and EVRL of ₹1230.90. The interest on working and fixed capital worked out to be ₹441.45 and ₹167.54, respectively. The overall average first year total establishment costs per acre were ₹5298.32 out of which ₹1620.69 were incurred on digging, filling and planting, ₹1113.77 on preparation of land, ₹350.60 on planting material, ₹93.58 on irrigation, ₹38.06 on training/pruning, ₹3.83 on plant protection and ₹260.33 on (manures + fertilizers), ₹126.74 as depreciation and EVRL of ₹1105.35. The per acre interest on working and fixed capital worked out to be ₹437.51 and ₹147.85, respectively.

The year wise establishment costs per acre of kinnow orchards are given in Table 4.21 and Figure 4.9. The total establishment cost per acre incurred on kinnow orchards were ₹12548.13 in case of marginal orchards, ₹13090.90 in case of small orchards and ₹13942.77 in case of medium orchards with an overall average of ₹12707.49. The data in the table further indicated that the first year establishment costs for marginal, small and medium orchards were ₹5262.05, ₹5391.00 and

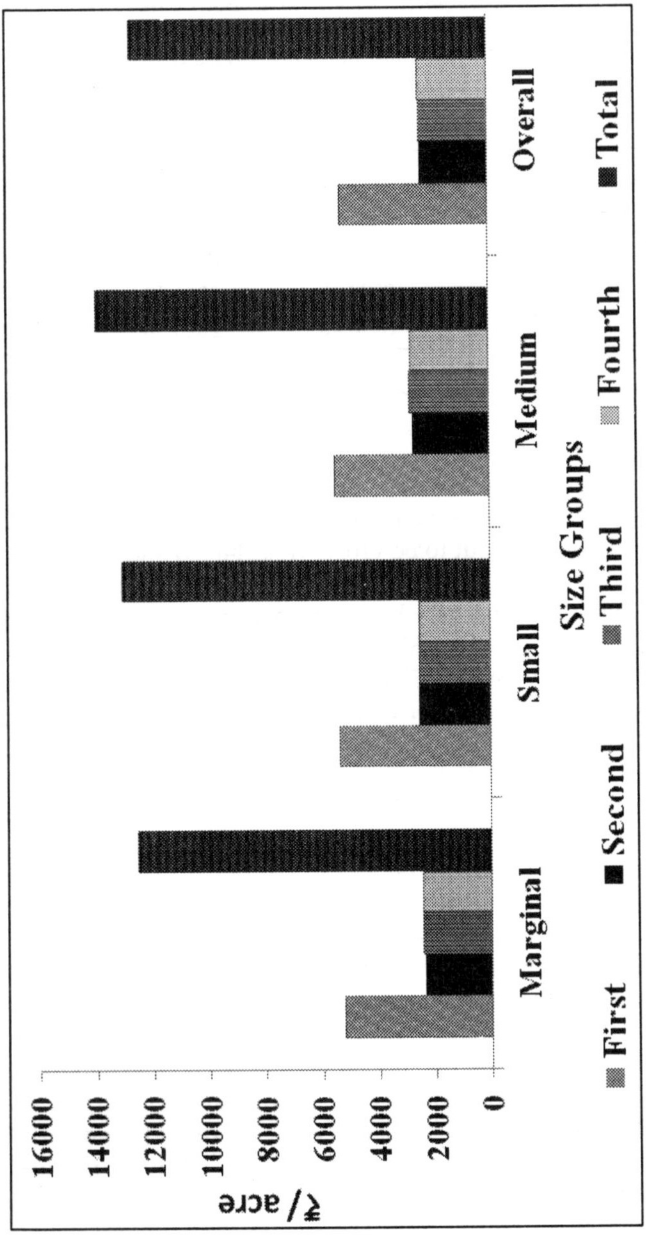

Figure 4.9: Year-wise Establishment Costs (₹/acre) Under Different Size Groups of Kinnow Orchards

₹5523.92, respectively, ₹2365.09, ₹2556.81 and ₹2750.12, respectively for the second year, ₹2445.31, ₹2569.86 and ₹2848.17, respectively for the third year whereas for the fourth year they were ₹2475.23, ₹2573.23 and ₹2820.56, respectively. The average, first, second, third and fourth year per acre establishment costs for all the size of holdings taken together were ₹5298.32, ₹2418.39, ₹2484.18 and ₹2506.60, respectively.

Table 4.21: Year-wise Establishment Costs Under Different Size Groups of Kinnow Orchards

(₹/acre)

Year	*Marginal*	*Small*	*Medium*	*Overall*
I	5262.50	5391.00	5523.92	5298.32
II	2365.09	2556.81	2750.12	2418.39
III	2445.31	2569.86	2848.17	2484.18
IV	2475.23	2573.23	2820.56	2506.60
Total	**12548.13**	**13090.90**	**13942.77**	**12707.49**

4.4.2.2.2 Operational Costs

The item wise and concept-wise operational costs are presented in Table 4.22 and Figure 4.10 and 4.11. The per acre cost of cultivation towards cost A, cost B and cost C, respectively worked out to be ₹1050.81, ₹2440.28 and ₹3690.08 in marginal orchards, ₹1430.33, ₹2880.57 and ₹3967.57 in small orchards and ₹1308.16, ₹2895.37 and ₹4091.37 in medium orchards, whereas per acre overall average cost worked out to be ₹1142.29, ₹2550.61 and ₹3762.89, respectively. In case of marginal orchards the per acre costs incurred were ₹710.85 on hired human labour, ₹93.15 on training/pruning), ₹117.62 on (manures + fertilizers), ₹16.60 on plant protection, ₹112.59 as interest on working capital, ₹123.50 as depreciation, ₹1093.69 as EVRL, ₹172.08 as interest on fixed capital and ₹1250.00 on family human labour. In case of small orchards, the per acre costs incurred were ₹803.70.85 on hired human labour, ₹139.64 on training/pruning), ₹294.74 on (manures + fertilizers), ₹39.0 on plant protection, ₹153.25 as interest on working capital, ₹142.25 as depreciation, ₹1129.00 as EVRL, ₹178.99 as interest on fixed capital and ₹1087.00 on family human labour. Similarly, in medium orchards the per acre costs incurred were ₹740.65 on hired human labour, ₹196.47 on training/pruning), ₹195.13 on (manures + fertilizers), ₹35.75 on plant protection, ₹140.16 as interest on working capital, ₹165.21 on depreciation, ₹1230.90 as EVRL, ₹191.10 as interest on fixed capital and ₹1196.00 on family human labour. The data further indicated that on an average the per acre expenditures involved were ₹732.31 on hired human labour, ₹1212.28 on family human labour, ₹159.13 on (manures+fertilizers), ₹1105.35 as earned value of rented land, ₹22.11 as plant protection, ₹106.35 as training/pruning), ₹122.39 as interest on working capital, ₹128.83 as depreciation and ₹174.14 as interest on fixed capital.

4.4.2.2.3 Returns

The per year per acre returns from kinnow orchards for different age groups (upto 10 years, 11th to 15th year and above 15th year) are presented in Table 4.23 and

Table 4.22: Item-wise and Concept-wise Operational Costs Under Different Size Groups of Kinnow Orchards

(₹/acre)

Sl.No.	*Item*	*Marginal*	*Small*	*Medium*	*Overall*
1	Hired Human labour	710.85	803.70	740.65	732.31
2	Irrigation	0.00	0.00	0.00	0.00
3	Training/Pruning	93.15	139.64	196.47	106.35
4	Manures + fertilizers	117.62	294.74	195.13	159.13
5	Plant protection	16.60	39.00	35.75	22.11
6	Interest on working capital	112.59	153.25	140.16	122.39
7	Land revenue	0.00	0.00	0.00	0.00
8	Depreciation	123.50	142.25	165.21	128.83
9	EVRL*	1093.69	1129.00	1230.90	1105.35
10	Interest on fixed capital	172.08	178.99	191.10	174.14
11	Family human labour	1250.00	1087.00	1196.00	1212.28
12	Cost A (1-6)	**1050.81**	**1430.33**	**1308.16**	**1142.29**
13	Cost B (1-10)	**2440.08**	**2880.57**	**2895.37**	**2550.61**
14	Cost C (1-11)	**3690.08**	**3967.57**	**4091.37**	**3762.89**

*: Earned value of rented land.

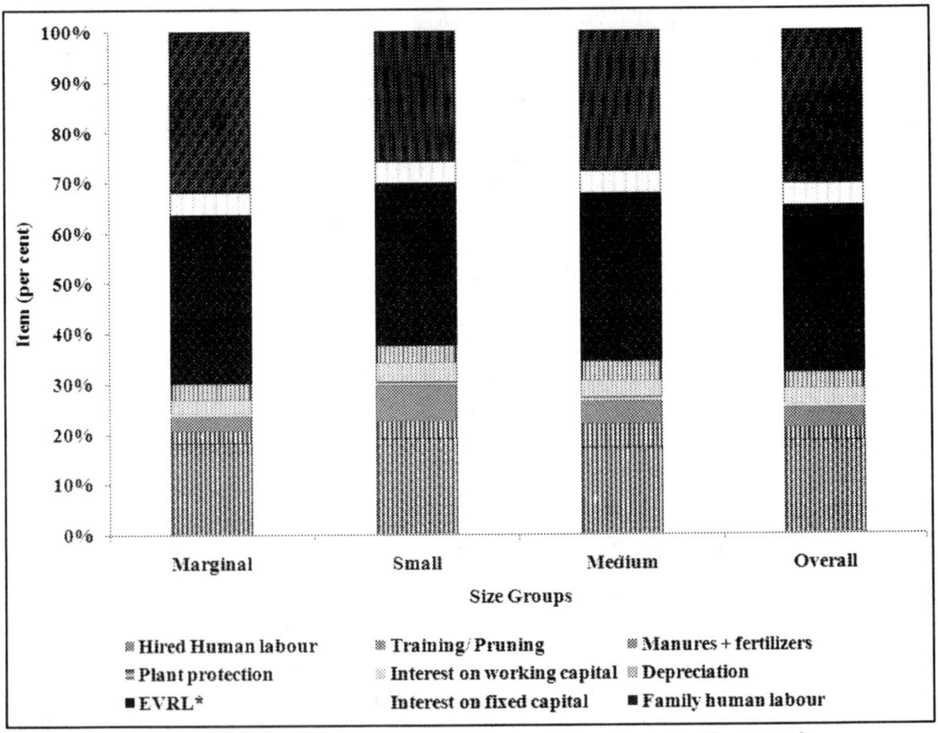

Figure 4.10: Item-wise Per Acre Operational Costs (Per cent) Under Different Size Groups of Kinnow Orchards

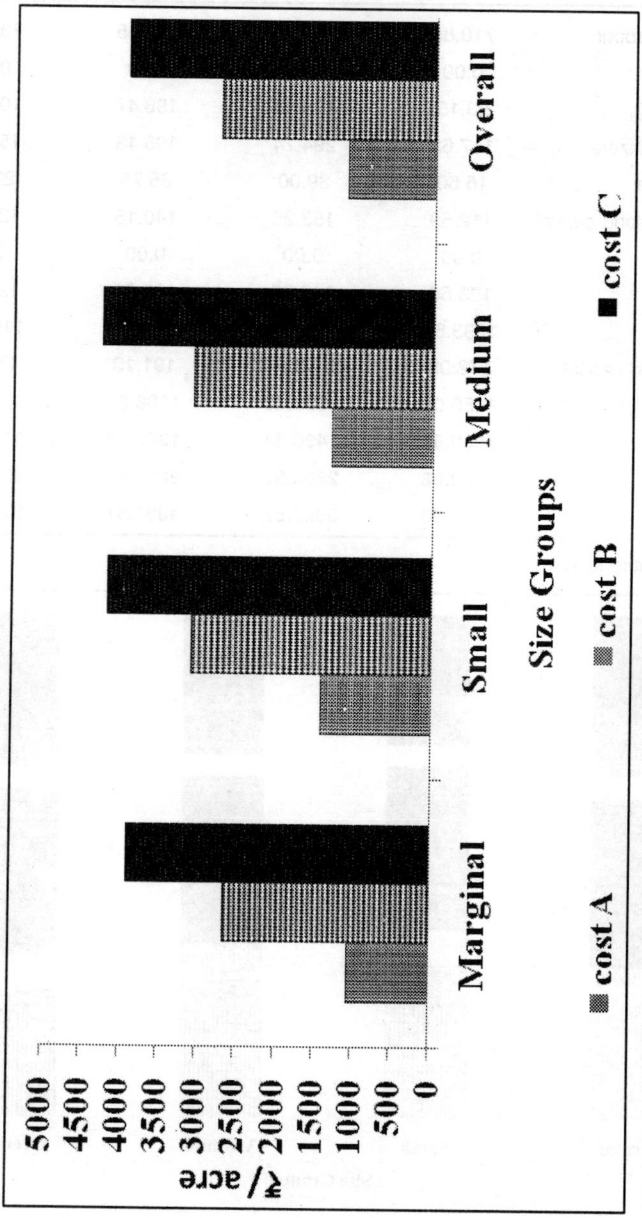

Figure 4.11: Concept-wise Cost of Production of Kinnow (₹/acre) Under Different Size Groups

Figure 4.12. The returns upto 10th year of age were ₹5526.30 in marginal orchards, ₹5862.65 in small orchards and ₹6235.55 in medium orchards. Similarly, it was ₹6785.32, ₹6954.35 and ₹7256.87 for marginal, small and medium orchards, respectively from 11th to 15th year while as from 15th year onwards the returns per acre per year were ₹6954.32 in marginal orchards, ₹7125.35 in small orchards and ₹7965.50 in medium orchards. The average returns per acre per year from marginal, small and medium orchards were ₹6562.11, ₹6774.05 and ₹7385.38, respectively with an overall average of ₹6632.07 for the whole life span of the kinnow orchard per year per acre.

4.4.2.3. Economic Viability

The economic viability of kinnow orchards is presented in Table 4.24. The net present value was worked out to be ₹7467.53 in marginal orchards, ₹9387.79 in small orchards and ₹11649.23 in medium orchards. The internal rate of return was 14.75 per cent in marginal orchards, 15.50 per cent in small orchards and 16.00 per cent in medium orchards. The benefit cost ratio was 1.65, 1.07 and 1.65 in marginal, small and medium orchards with pay-back period of 7.5, 7.8 and 7.2, respectively. On an average the net present value of kinnow orchards was ₹7929.39 having internal rate of return 15.42 per cent, benefit cost ratio 1.52 and having a pay-back period of 7.6 years.

Table 4.23: Average Returns Under Different Age Groups of Kinnow Orchards

(₹/acre/year)

Item	*Marginal*	*Small*	*Medium*	*Overall*
Upto 10 years	5526.30	5862.65	6235.55	5620.75
11th to 15th year	6785.32	6954.35	7256.87	6835.98
Above 15th year	6954.32	7125.35	7965.50	7020.42
Overall	**6562.11**	**6774.05**	**7385.38**	**6632.07**

Table 4.24: Economic Viability Under Different Size Groups of Kinnow Orchards

Group	*Net Present Value (₹)*	*Internal Rate of Return (per cent)*	*Payback Period (Years)*	*Benefit Cost Ratio*
Marginal	7467.53	14.75	7.5	1.65
Small	9387.79	15.50	7.8	1.07
Medium	11649.23	16.00	7.2	1.65
Overall	**7929.39**	**15.42**	**7.6**	**1.52**

4.4.3 Economics of Production of Lemon

4.4.3.1 Resource Use Efficiency

The Cobb-Douglas production function was fitted on the similar pattern as was done in case of orange and kinnow and grouping of orchards was also followed on

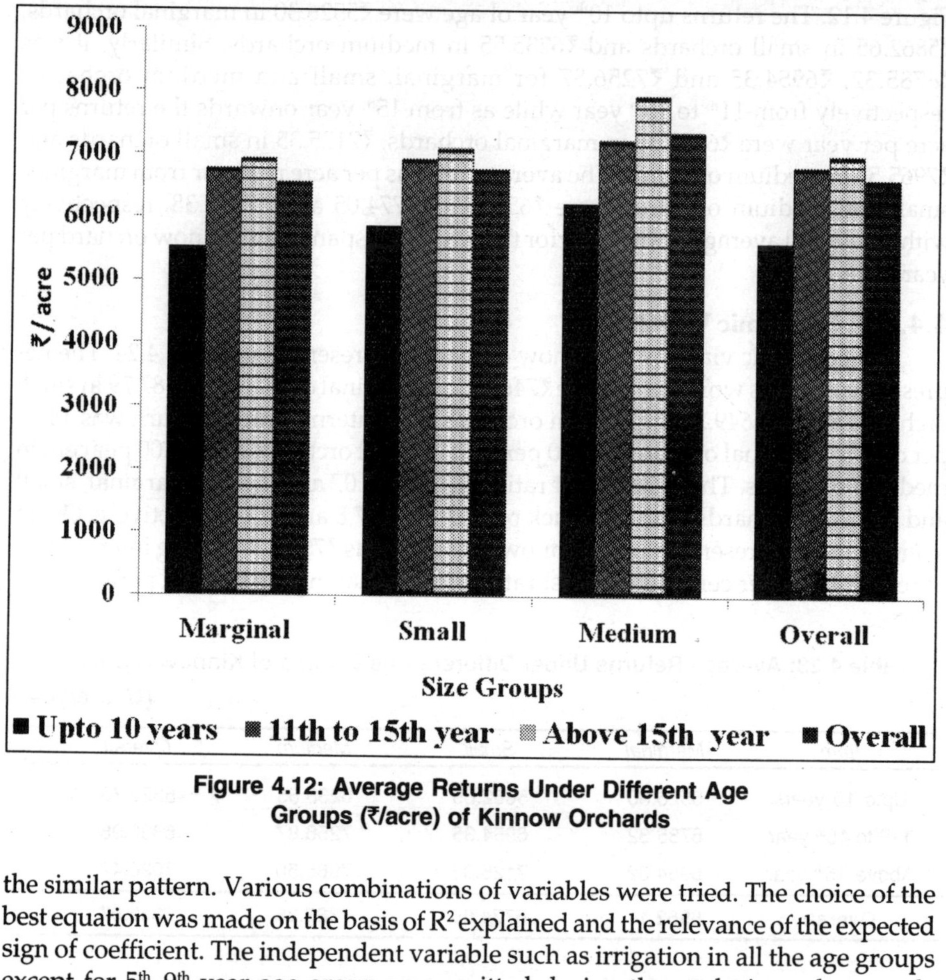

Figure 4.12: Average Returns Under Different Age Groups (₹/acre) of Kinnow Orchards

the similar pattern. Various combinations of variables were tried. The choice of the best equation was made on the basis of R^2 explained and the relevance of the expected sign of coefficient. The independent variable such as irrigation in all the age groups except for 5th–9th year age group was omitted during the analysis as the sample orchardists had not used this resource in other age groups in the selected area.

The regression function result and marginal value productivity of lemon from 5th–9th year is given in Table 4.25. The output of lemon orchards was regressed against human labour, (manures + fertilizers), irrigation, plant protection and training/pruning. The perusal of the data revealed that production function from 5th–9th year was statistically significant having R^2 value (0.729) meaning that 72.9 per cent of the total variations in the production function for lemon was explained by the explanatory variables under consideration. The functional analysis for lemon production revealed that human labour and irrigation were positively and negatively significant at 1 per cent level of probability, respectively with regression coefficients as 1.300 and -0.062, respectively whereas (manures + fertilizers) with negative non-significance had the regression coefficient as -0.845, plant protection and training/pruning were positively non significant with their regression values as 0.012 and 0.020, respectively. The marginal value productivity of human labour, plant protection and training/pruning

was positive with their values at 0.103, 0.016 and 0.040, respectively, whereas (manures + fertilizers) and irrigation were negative with their values as -0.482 and -0.032, respectively.

Table 4.25: Estimated Regression Coefficients of Various Factors, their Standard Errors and MVP of Lemon Production (5th–9th Year)

Variables	*Regression Coefficients*	*Standard Error*	*MVP*
Constant	-0.083	1.262	
Manures + Fertilizers	-0.845	0.354	-0.482
Irrigation	-0.062*	0.015	-0.032
Plant Protection	0.012	0.011	0.016
Training/Pruning	0.020	0.011	0.040
Human Labour	1.300*	0.338	0.103
F value	4.84		
Coefficient of determination (R^2)	**0.729***		

Note: * Significant at 1 per cent level of significance.

The regression function and marginal value productivity of lemon from 10th – 14th year is depicted in Table 4.26, wherein the output of lemon orchards was regressed against human labour, (manures + fertilizers), plant protection and training/pruning. The production function with R^2 value as 0.830 was statistically significant meaning that 83.0 per cent of the variations in the production of lemon was explained by the explanatory variables under consideration. The table further indicated that the human labour with statistically positive significance at 1 per cent level of probability had the regression value of 1.543 whereas (manures + fertilizers) had the regression value of -0.096 and was negatively significant at 1 per cent level of probability. The marginal value productivity for human labour alone was positive (0.179) whereas rest of the variables was negative *i.e.*, -0.054 for (manures + fertilizers), -0.007 for plant protection and -0.046 for training/pruning.

Table 4.26: Estimated Regression Coefficients of Various Factors, their Standard Errors and MVP of Lemon Production (10th–14th Year)

Variables	*Regression Coefficients*	*Standard Error*	*MVP*
Constant	-2.670**	0.856	
Manures + Fertilizers	-0.096*	0.027	-0.054
Plant Protection	-0.029	0.015	-0.007
Training/Pruning	-0.014	0.011	-0.046
Human Labour	1.543*	0.270	0.179
F value	12.16		
Coefficient of determination (R^2)	**0.830***		

Note: * Significant at 1 per cent level of significance; **: Significant at 5 per cent level of significance.

The regression function result and marginal value productivity of lemon from 15th – 19th year presented in Table 4.27 depicted that the output of lemon orchards was regressed against human labour, (manures + fertilizers) and plant protection and the production function used was statistically significant having R^2 value as high as 0.950 meaning that 95.0 per cent of the total variation in the production for lemon was explained by these variables. The functional analysis for lemon production revealed that manures + fertilizers with significance at 1 per cent level of probability had regression coefficient as 0.972, whereas plant protection was found to be significant at 5 per cent level of probability with regression coefficient as -0.079 but human labour was non significant with value of regression coefficient as 0.053. The marginal value productivity of human labour and manures + fertilizers was positive with their values as 0.066 and 0.075, respectively whereas for plant protection it was negative with its value as -0.005.

Table 4.27: Estimated Regression Coefficients of Various Factors, their Standard Errors and MVP of Lemon Production (15th–19th year)

Variables	*Regression Coefficients*	*Standard Error*	*MVP*
Constant	-0.450**	0.202	
Manures + Fertilizers	0.972*	0.051	0.075
Plant Protection	-0.079**	0.032	-0.005
Human Labour	0.053	0.044	0.066
F value	238.37		
Coefficient of determination (R^2)	**0.950****		

Note: * Significant at 1 per cent level of significance; **: Significant at 5 per cent level of significance.

The regression function result and marginal value productivity of lemon orchards for the overall period presented in Table 4.28 depicted that the output of lemon orchards was regressed against human labour, (manures + fertilizers), irrigation, plant protection and training/pruning. The production function used was statistically significant having R^2 value as 0.815 meaning that 81.5 per cent of the total variations in the production for lemon was explained by the explanatory variables under consideration. The functional analysis for lemon production revealed that human labour was found to be positively significant at 5 per cent level of probability with regression coefficient as 0.451 while (manures + fertilizers) was found to be positively significant at 1 per cent level of probability with regression coefficient as 1.257. The variables like irrigation, plant protection and training/pruning were found to be negative and non significant with regression coefficient as –0.011, –0.002 and –0.023, respectively. The marginal value productivity of human labour and (manures + fertilizers) was positive with their values as 0.111 and 0.882, respectively whereas that of irrigation, plant protection and training/pruning was negative with their values as -0.020, -59.710 and -0.039, respectively.

Table 4.28: Estimated Regression Coefficients of Various Factors, their Standard Errors and MVP of Lemon Production (Overall)

Variables	*Regression Coefficients*	*Standard Error*	*MVP*
Constant	-2.744**	1.016	
Manures + Fertilizers	1.257*	0.330	0.882
Irrigation	- 0.011	0.016	-0.020
Plant Protection	-0.002	0.014	-59.710
Training/Pruning	-0.023	0.011	-0.039
Human Labour	0.451**	0.194	0.111
F value	27.61		
Coefficient of determination (R^2)	**0.815****		

Note: * Significant at 1 per cent level of significance; **: Significant at 5 per cent level of significance.

4.4.3.2. Costs and Returns

Lemon being a perennial fruit crop starts bearing three or four years after plantation. During the productive life of the tree, it continues to bear fruits and yields sizeable income to the growers. The information in respect of yearly costs involved in the establishment of lemon orchards was necessary in order to allocate the non recurring cost to the yearly cost of production.

4.4.3.2.1 Establishment Cost

Like orange and kinnow, lemon also requires a high investment for its establishment during the first year in order to perform various operations. After the first year relatively low cost is incurred till the bearing of the orchard. The operation wise first year per acre establishment costs of lemon is presented in Table 4.29 and Figure 4.13. Since the number of orchardists involved in lemon cultivation were less, therefore, it was not possible to find out the results for different group of holdings *viz.* marginal, small, medium and large. So, an overall costs and returns estimation was made. The first year per acre establishment costs incurred were ₹644.60 on preparation of land, ₹1028.06 on digging, filling and planting, ₹221.54 on planting material, ₹20.00 on training/pruning, ₹67.78 on irrigation, ₹150.48 on (manures + fertilizers), ₹231.00 as depreciation on machinery and farm inventory and ₹1050.82 as earned value of rented land (EVRL). The interest on working and fixed capital worked out to be ₹253.49 and ₹153.82, respectively. The total first year cost involved in establishing a lemon orchard was ₹3821.59 per acre.

Table 4.30 and Figure 4.16 presented the year-wise per acre establishment costs of lemon orchards. The data in the table indicated that the total establishment costs per acre incurred on lemon orchards were ₹9563.54. Whereas first year establishment costs were ₹3821.59, second year, it was ₹1852.66, while during the third year and fourth year it was ₹1893.54 and ₹1995.75, respectively.

4.4.3.2.2 Operational Costs

The item-wise and concept wise operational costs are presented in Table 4.29 and Figure 4.14 and 4.15. The per acre cost of cultivation towards cost A, cost B and

cost C, respectively worked out to be ₹816.77, ₹2241.64 and ₹2930.27. The table further revealed that the per acre costs incurred on hired human labour were ₹597.16, ₹12.39 on training/pruning, ₹99.34 on (manures + fertilizers) and ₹11.44 on plant protection. The cost involvement for interest on working capital was ₹96.44, ₹120.91 on depreciation, ₹1050.82 on EVRL, ₹253.14 as interest on fixed capital and ₹688.63 on family human labour.

Table 4.29: First Year Establishment Costs and Item-wise as well as Concept-wise Operational Costs of Lemon Orchards

(₹/acre)

Sl.No.	*Operations*	*Establishment costs (1st year)*	*Operational cost*
1.	Preparation of land	644.60	–
2.	Digging, filling and planting	1028.06	–
3.	Planting material	221.54	–
4.	Hired human labour	–	597.16
5.	Irrigation	67.78	0.00
6.	Training/Pruning	20.00	12.39
7.	Manures + fertilizers	150.48	99.34
8.	Plant protection	0.00	11.44
9.	Interest on working capital	253.49	96.44
10.	Land revenue	0.00	0.00
11.	Depreciation	231.00	120.91
12.	Earned value of rented land	1050.82	1050.82
13.	Interest on fixed capital	153.82	253.14
14.	Family human labour	–	688.63
15.	Cost A (4-9)	–	**816.77**
16.	Cost B (4-13)	–	**2241.64**
17.	Cost C (4-14)	–	**2930.27**
	Total	**3821.59**	–

Table 4.30: Year-wise Establishment Costs and Average Returns Under Different Age Groups of Lemon Orchards

(₹/acre)

Age Groups (in Years)	*Establishment Costs*	*Average Returns*
0–1	3821.59	–
1–2	1852.66	–
2–3	1893.54	–
3–4	1995.75	–
5–10	–	6396.15
11–15	–	14178.85
Above 15th year	–	12984.62
Total	**9563.54**	**10475.36**

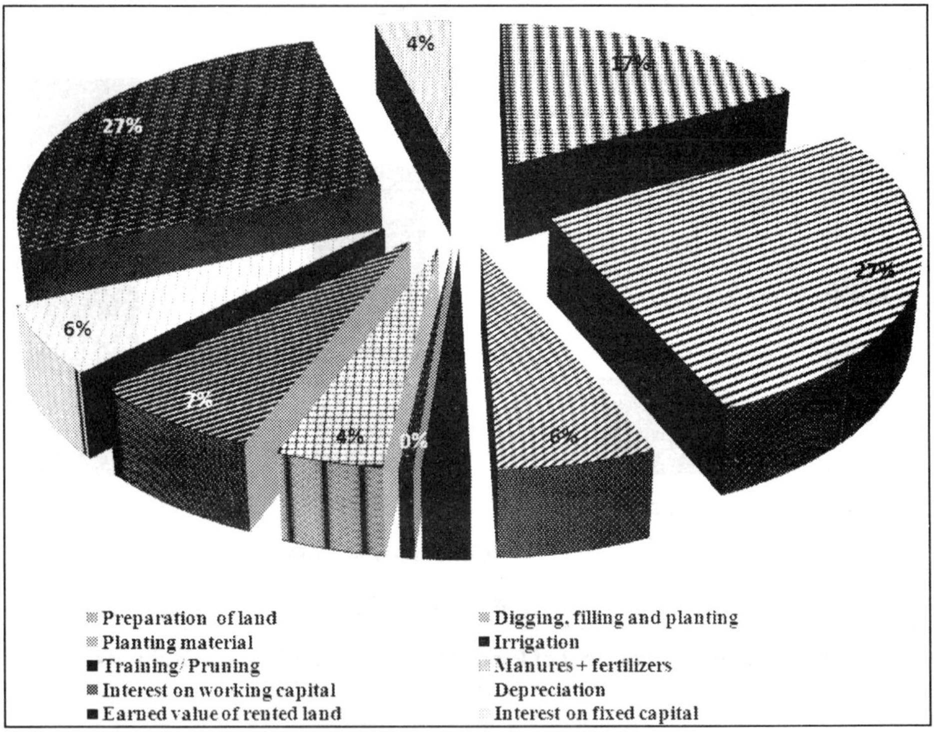

Figure 4.13: Per cent Establishment Costs of Lemon Orchards during 1st Year

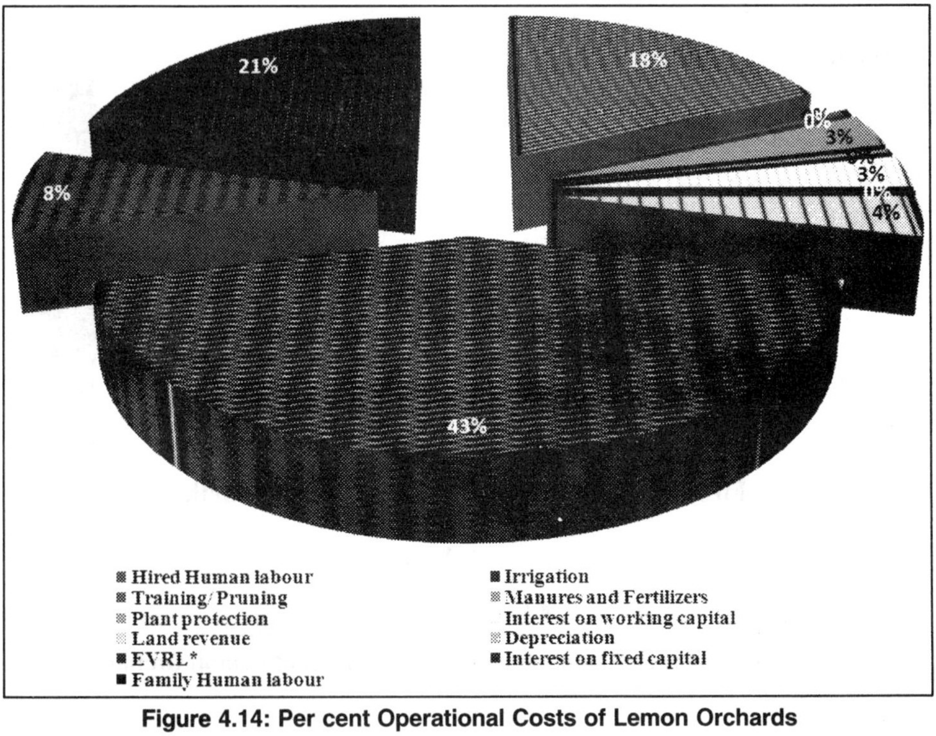

Figure 4.14: Per cent Operational Costs of Lemon Orchards

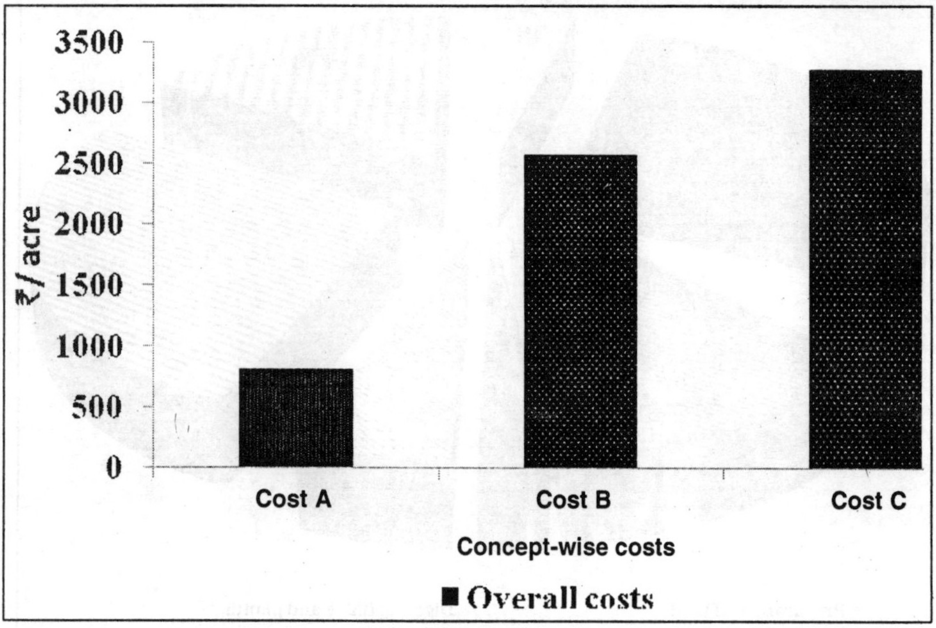

Figure 4.15: Pattern of Cost of Production of Lemon (₹/acre) Orchards

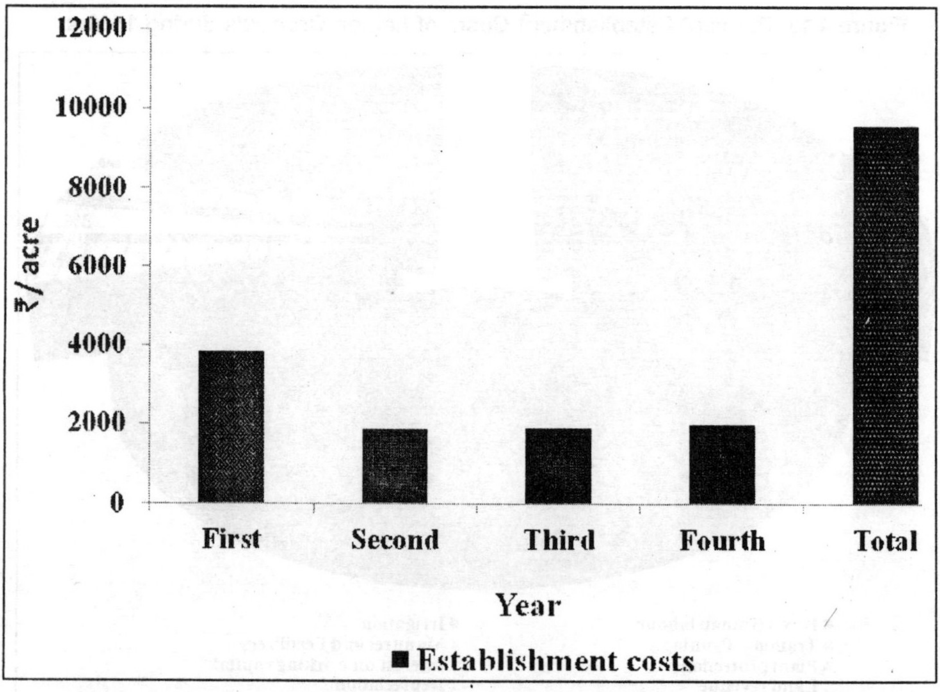

Figure 4.16: Year-wise Establishment Costs (₹/acre) of Lemon Orchards

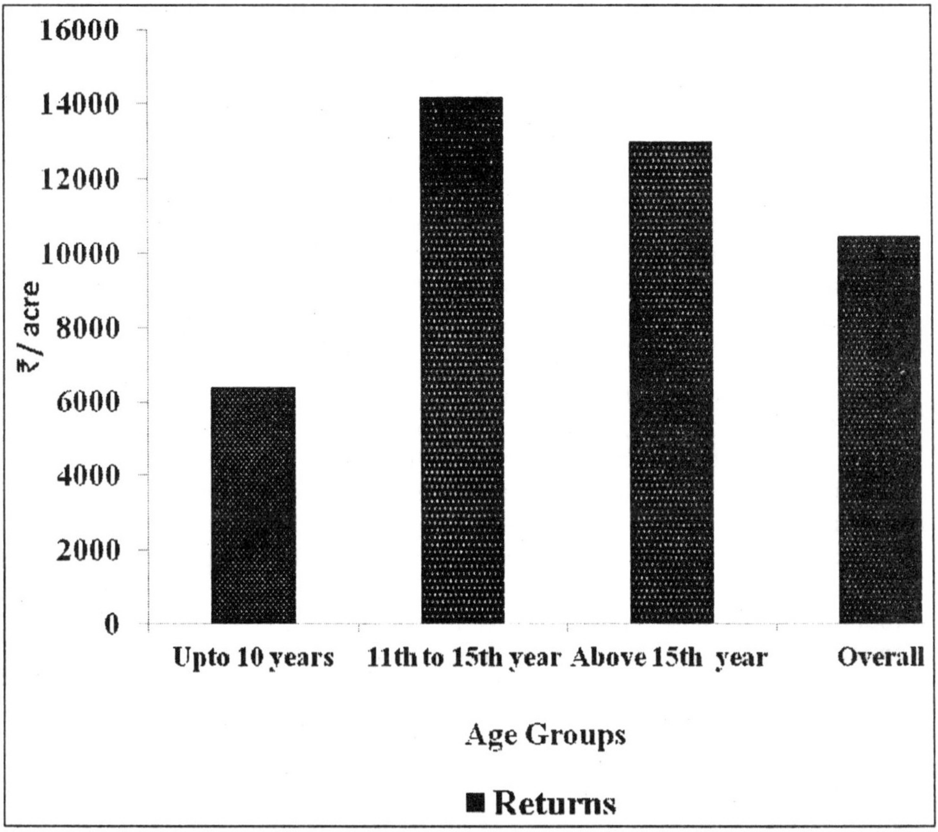

Figure 4.17: Average Returns Under Different Age Groups (₹/acre) of Lemon Orchards

4.4.3.2.3 Returns

The returns per year per acre from lemon orchards were worked out for different age groups as presented in Table 4.30 and Figure 4.17. The returns upto 10th year of age were ₹6396.15, from 11th to 15th year ₹14178.85, whereas from 15th year onwards the returns per acre per year were ₹12984.62. On an average, the returns per acre per year from lemon orchards were ₹10475.36.

4.4.3.3. Economic Viability

The economic viability of lemon orchards is presented in Table 4.31. The net present value was worked out to ₹5475.61, the internal rate of return was 20.80 per cent, the benefit cost ratio was 2.70 with pay-back period of 6.4 years.

Table 4.31: Economic Viability of Lemon Orchards

Fruit	*Net Present Value (₹)*	*Internal Rate of Return (per cent)*	*Payback Period (Years)*	*Benefit Cost Ratio*
Lemon	5475.61	20.80	6.4	2.70

4.4.4. Marketing of Citrus

In planned economic development programme, exchange of goods play a very important role in maintaining equilibrium between production and consumption. The marketing of agricultural products is becoming more important as created by the new world trade order under the WTO agreements. The small and resource poor cultivators have to face big world players for marketing their produce. The prosperity of the cultivators thus depends not only on the increased rate of production, but also upon the method and efficiency with which they dispose of their produce to the great advantage. It assumes special significance in the marketing of perishable commodities like citrus because very small portion of it is consumed by the farm families, therefore, farmers have more marketable surplus.

In specialized farming the producers who are in a position to adjust their production to the demand, reap the maximum benefit of the market. If cultivators are unable to adjust their production to the demand of the market, there can be no appreciable improvement in their condition even if the output is better in quality and larger in quantity. Inspite of the partial failure to adjust the production on the farm to the demand of the market, the progressive farmers realize the importance of the study of the market. The element of time is an important factor in marketing of agricultural produce in general and fruits in particular. The marketing possibility of the perishable commodities like fruits depends very largely on the rapidity with which they can be transported to the market. An efficient system of marketing, however, would require many aspects like less number of intermediaries, nominal commission, loading/ unloading charges, minimum marketing cost besides the development of means of transport. Efficient marketing should be such that the produce should reach the consumer in good state without damage, with less cost and within a short time after the produce is harvested.

Market Structure of Citrus Marketing

It was imperative to study the activities of each agency taking part in citrus trade in the study area. The agencies that facilitated the flow of citrus fruits till they reached the ultimate consumer were commission/forwarding agents, wholesalers and retailers.

(a) Commission Agent/Forwarding Agent

The function of the commission agent is to sell the produce of a producer without any risk of loss or cost in the study area. In lieu of his services, he charges certain percentage on the total sale value of the commodity. For marketing of selected citrus fruits sold in various markets, the commission agents charged 5 per cent of sale value and sometimes more than 5 per cent. Many commission agents also performed the function of wholesalers and therefore obtained maximum profit out of citrus trade.

(b) Wholesaler

Wholesalers have got the key position in fruit marketing and also play sometimes the role of commission agent as well as broker. Generally, they purchased the produce either from commission agents or directly from the producers in the market.

(c) Retailer

The retailer is a marketing functionary, who caters the needs of the consumers by retailing, generally keep a small establishment and reap a maximum profit especially in fruit trade due to their higher share in the consumers' rupee. Mostly they buy the produce from wholesalers as well as producers early in the morning and sell it out during the remaining day and thus a vertically integrated markets benefit both the producers and consumers. Some small producers themselves perform the function of retailing and receive better price of their produce in the nearby local markets. Owing to these facts, a survey of the selected farmers in marketing their produce was conducted and the results obtained are presented under various heads.

4.4.4.1. Marketing Channels in Selected Study Area

The chain of various intermediaries/functionaries commonly known as marketing channel comprising of agencies like producers, commission/forwarding agent, wholesalers and retailers etc. and also sometimes, direct sale of produce help in distribution of fruits/crops from producers to ultimate consumers in Jammu region. The marketing channel operating under Jammu conditions are as under:

I. Producer → Forwarding/Commission agent → Retailer → Consumer

II. Producer → Wholesaler → Retailer → Consumer

III. Producer → Retailer → Consumer

IV. Producer → Consumer

The flow chart 4.18 depicts the path along which, the citrus fruit grown in various parts of Jammu region flows with subsidiary routes from the producer to ultimate consumer. It has been found that in study area of the Jammu region, forwarding/ commission agents play a predominant role in marketing of fruits. Sometimes, producer sells directly to the retailers or consumers and sometimes they do marketing through wholesalers. In general, the channels of distribution of citrus fruit in various districts of Jammu region are almost one and the same, whether it is kinnow, orange or lemon.

Table 4.32: Quantity of Citrus Fruit Sold Through Different Marketing Channels in Various Districts of Jammu Region

(Quintals)

Marketing Channels	*Quantity Sold*				
	Jammu	*Rajouri*	*Kathua*	*Samba*	*Jammu region*
I	315.00 (49.13)	295.49 (45.85)	714.00 (38.95)	856.00 (62.77)	2180.49 (48.65)
II	124.00 (19.34)	108.29 (16.80)	430.00 (23.46)	Nil	662.29 (14.78)
III	147.80 (23.05)	150.69 (23.38)	561.95 (30.66)	404.15 (29.64)	1264.59 (28.21)
IV	54.40 (8.48)	90.00 (13.96)	126.95 (6.93)	103.46 (7.59)	374.81 (8.36)
Total	**641.20 (100.00)**	**644.47 (100.00)**	**1832.90 (100.00)**	**1363.61 (100.00)**	**4482.18 (100.00)**

Figure in parentheses are the percentage of total.

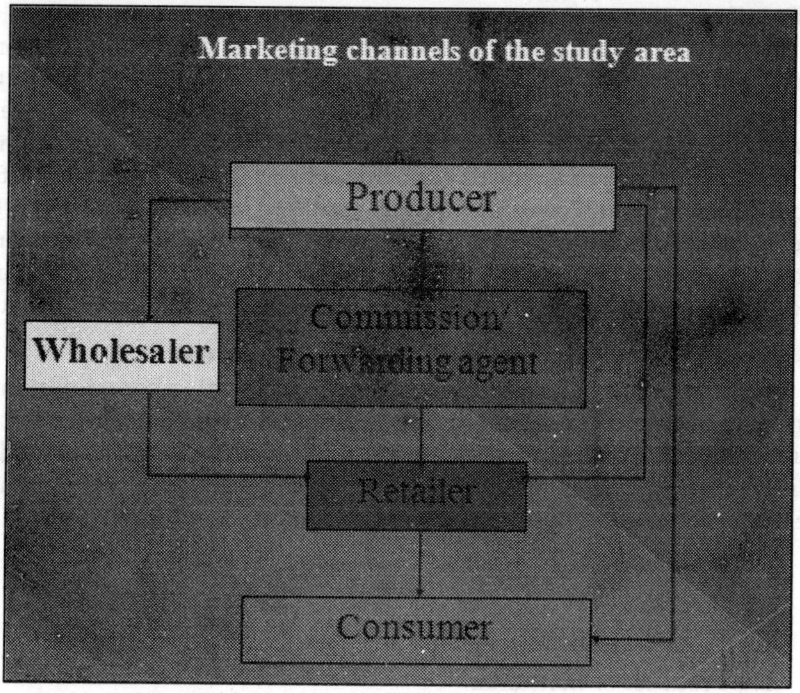

Figure 4.18: Flow Chart of Various Marketing Channels of the Study Area

The quantity of citrus sold through the different marketing channels is given in Table 4.32. The Table revealed that the quantity sold through channel I, II, III and IV in Jammu district was worked out to be 315.00, 124.00, 147.80 and 54.40 quintals, respectively, in Rajouri district 295.49, 108.29, 150.69 and 90.00 quintals, respectively, in Kathua district 714.00, 430.00, 561.95, 126.95 quintals, respectively and in Samba district 856.00 quintals through channel I, 404.15 quintals through channel III and 103.46 quintals through channel IV were sold. The table further revealed that out of the total citrus marketed in Jammu region, about 2180.49 quintals was sold through channel I *i.e.*, through commission/forwarding agent, 662.29 quintals through channel II *i.e.*, through wholesaler, 1264.59 quintals through channel III *i.e.*, through retailer and 374.81 quintals through channel IV *i.e.*, directly to consumer.

4.4.4.2. Marketing Costs, Margins, Losses and Price Spread

4.4.4.2.1 Marketing Cost of Citrus

In general marketing costs constitute the expenses on the items like picking, filling, packing, transportation, loading and unloading, commission and other charges. These costs are the actual expenditure incurred for the smooth running of business as well as for efficient marketing of particular farm commodity.

The channel wise decomposition of marketing costs components for citrus fruit in Jammu district is given in Table 4.33. The table revealed that the major items of marketing expenses in all the channels at producers' level included picking, filling, depreciation of container, transportation cost, loading and unloading charges and

commission of the forwarding/commission agent. These costs varied to the extent of ₹760.00, ₹552.50, ₹425.71 and ₹304.00 per quintal for channel I, II, III and IV, respectively, where as it was ₹112.27, ₹111.25, ₹113.57 and ₹111.00, respectively for picking and filling. It was further observed from the table that for loading/unloading of fruits, the cost incurred was ₹10.00 per quintal for each and every channel. The transportation cost amounted to ₹320.23, ₹231.25, ₹152.14 and ₹103.00 for channel I, II, III and IV, respectively. The per quintal miscellaneous charges were found to be highest in channel IV (₹8.50) followed by channel II (₹3.75) whereas it was ₹3.50 each in channel I and III. An expenditure of ₹67.50 per quintal was the commission of the commission/forwarding agent in channel I where as it was nil for the wholesaler and retailer as is clear from the table. Marketing cost borne by the wholesaler in channel II was ₹67.50 per quintal in which the transportation cost amounted to ₹57.50 per quintal followed by loading/unloading charges of ₹10.00 per quintal. At the retailers' level, transportation cost, loading and unloading, shop/rehri charges and cost of plastic bags were the major items of marketing cost in channel I and II where as in case of channel III transportation cost was not borne by the retailer. The perusal of the data further indicated that in channel IV whole of the marketing cost was borne by the producer as there was the direct marketing of produce. As far as wholesalers' and retailers' marketing cost was concerned, in channel I, retailer had incurred the cost of ₹30.49 while it was nil for wholesaler. In channel II, the retailer incurred the cost of ₹29.77 while it was ₹67.50 for the wholesaler. In channel III again, the retailer had incurred the cost of ₹18.50, but for wholesaler, it was zero.

The channel wise decomposition of marketing costs components for citrus fruit in Rajouri district is given in Table 4.34. The major items which played important role in marketing cost in all the channels at producers' level included picking, filling, depreciation of container, transportation cost, loading and unloading charges and traders' commission. These costs varied to the extent of ₹396.00, ₹250.00, ₹242.05 and ₹214.00 per quintal for channel I, II, III and IV, respectively with ₹110.00 for each channel as far as picking and filling was concerned and for loading/unloading of fruits the producer incurred a cost of ₹10.00 per quintal for each and every channel. The transportation cost for various channels (I, II, III and IV) amounted to ₹73.20, ₹60.00, ₹50.23 and ₹40.00, respectively. The table further revealed that the grower also incurred expenditure on depreciation of container which ranged between ₹47.00 and ₹138.00. Here, in channel I where commission agent was involved, he usually charged ₹62.80 per quintal (5 per cent of the value of produce). The miscellaneous charges were found to be highest in channel IV (₹7.00 per quintal) followed by channel II (₹2.50 per quintal), channel I (₹2.00 per quintal) and channel III (₹1.50 per quintal). In channel-II, an estimated expenditure of ₹55.00 per quintal at wholesale level was incurred by the wholesaler with the maximum of ₹45.00 per quintal as the transportation cost. The retailer incurred the cost of ₹6.43 and ₹10.00 towards transportation charges in channel-I and II, respectively. It was also observed from the table that an amount of ₹12.75, ₹10.00 and ₹5.00 was incurred on loading and unloading in channel I, II and III, respectively and ₹4.10 in channel I and ₹4.50 each in channel II and III were paid for shop/rehri charges. Thus, on an average, the retailer incurred ₹33.28, ₹34.50 and ₹19.50 respectively as the marketing cost in

channel I, II and III, respectively. The total marketing cost involved in marketing of citrus in the selected sample area was ₹429.28 for channel I, ₹339.50 for channel II, ₹261.55 for channel III and ₹214.00 for channel IV.

Table 4.33: Channel-wise Decomposition of Marketing Costs Components for Citrus Fruit in Jammu District

(₹/q)

Sl.No.	*Functionary*	*Channel–I*	*Channel–II*	*Channel–III*	*Channel–IV*
1.	Marketing cost incurred by the producer	**760.00** (96.14)	**552.50** (82.65)	**425.71** (95.84)	**304.00** (100.00)
	i) Picking, Filling	112.27 (14.20)	111.25 (16.64)	113.57 (25.57)	111.00 (36.51)
	ii) Depreciation of container (Tokri/Crate/Gunny bags)	246.50 (31.18)	196.25 (30.20)	146.50 (32.98)	71.50 (23.52)
	iii) Transportation cost	320.23 (40.51)	231.25 (34.59)	152.14 (34.25)	103.00 (33.88)
	iv) Loading/unloading charges	10.00 (1.27)	10.00 (1.50)	10.00 (2.25)	10.00 (3.29)
	v) Miscellaneous charges	3.50 (0.44)	3.75 (0.58)	3.50 (0.79)	8.50 (2.80)
	vi) Commission	67.50 (8.54)	0.00 (0.00)	0.00 (0.00)	0.00 (0.00)
2.	Marketing cost incurred by the wholesaler	**0.00** (0.00)	**67.50** (10.09)	**0.00** (0.00)	**0.00** (0.00)
	i) Loading/unloading charges	0.00 (0.00)	10.00 (1.50)	0.00 (0.00)	0.00 (0.00)
	ii) Transportation cost	0.00 (0.00)	57.50 (8.60)	0.00 (0.00)	0.00 (0.00)
3.	Marketing cost incurred by the retailer	**30.49** (3.86)	**29.77** (4.45)	**18.50** (4.16)	**0.00** (0.00)
	i) Transportation cost	6.14 (0.78)	5.63 (0.84)	0.00 (0.00)	0.00 (0.00)
	ii) Loading/unloading charges	10.00 (1.27)	10.00 (1.50)	5.00 (1.13)	0.00 (0.00)
	iii) Shop/Rehri charges	4.35 (0.55)	4.14 (0.62)	3.50 (0.79)	0.00 (0.00)
	iv) Cost of plastic bags	10.00 (1.27)	10.00 (1.50)	10.00 (2.25)	0.00 (0.00)
	Total Marketing Cost (1+2+3)	**790.49**	**649.77**	**444.21**	**304.00**

Figure in parentheses are the percentage of total marketing cost of their respective channels.

The channel wise decomposition of marketing costs components for citrus fruit in Kathua district is depicted in Table 4.35. The marketing cost incurred by the producer was the highest with ₹ 427.40, ₹ 455.00, ₹285.75 and ₹ 261.67 per quintal for channel I, II, III and IV, respectively and lowest by the wholesaler with ₹ 60.00 in channel II.

Table 4.34: Channel-wise Decomposition of Marketing Costs Components for Citrus Fruit in Rajouri District

(₹/q)

Sl.No.	Functionary	Channel–I	Channel–II	Channel–III	Channel–IV
1	Marketing cost incurred by the producer	**396.00** (92.24)	**250.00** (73.64)	**242.05** (92.54)	**214.00** (100.00)
	i) Picking, Filling	110.00 (25.62)	110.00 (32.40)	110.00 (42.06)	110.00 (48.67)
	ii) Depreciation of container (Tokri/Crate/Gunny bags)	138.00 (32.15)	67.50 (19.88)	70.32 (26.89)	47.00 (21.96)
	iii) Transportation cost	73.20 (17.05)	60.00 (17.67)	50.23 (19.20)	40.00 (18.69)
	iv) Loading/unloading charges	10.00 (2.33)	10.00 (2.95)	10.00 (3.82)	10.00 (4.42)
	v) Miscellaneous charges	2.00 (0.47)	2.50 (0.74)	1.50 (0.57)	7.00 (3.27)
	vi) Commission	62.80 (14.63)	0.00 (0.00)	0.00 (0.00)	0.00 (0.00)
2.	Marketing cost incurred by the wholesaler	**0.00** (0.00)	**55.00** (16.20)	**0.00** (0.00)	**0.00** (0.00)
	i) Loading/unloading charges	0.00 (0.00)	10.00 (2.95)	0.00 (0.00)	0.00 (0.00)
	ii) Transportation cost	0.00 (0.00)	45.00 (13.25)	0.00 (0.00)	0.00 (0.00)
3.	Marketing cost incurred by the retailer	**33.28** (7.74)	**34.50** (10.16)	**19.50** (7.46)	**0.00** (0.00)
	i) Transportation cost	6.43 (1.50)	10.00 (2.95)	0.00 (0.00)	0.00 (0.00)
	ii) Loading/unloading charges	12.75 (2.97)	10.00 (2.95)	5.00 (1.91)	0.00 (0.00)
	iii) Shop/Rehri charges	4.10 (0.95)	4.50 (1.33)	4.50 (1.72)	0.00 (0.00)
	iv) Cost of plastic bags	10.00 (2.33)	10.00 (2.95)	10.00 (3.82)	0.00 (0.00)
	Total Marketing Cost (1+2+3)	**429.28**	**339.50**	**261.55**	**214.00**

Figure in parentheses are the percentage of total marketing cost of their respective channels.

The grower on an average incurred expenditure for picking and filling of fruits around ₹111.58, ₹110.00, ₹110.75 and ₹110.00 for channel I, II, III and IV, respectively while as for loading/unloading of fruits a cost of ₹10.00 per quintal in each channel was incurred. The transportation cost at the grower level amounted to ₹80.26, ₹101.67, ₹90.00 and ₹85.00 for channel I, II, III and IV, respectively. The data from the table further indicated that grower also incurred expenditure on depreciation of container which was highest in case of channel-II (₹231.00) followed by channel-I (₹151.28),

Table 4.35: Channel-wise Decomposition of Marketing Costs Components for Citrus Fruit in Kathua District

(₹/q)

Sl.No.	Functionary	Channel–I	Channel–II	Channel–III	Channel–IV
1.	Marketing cost incurred by the producer	**427.40** (93.26)	**455.00** (80.58)	**285.75** (93.80)	**261.67** (100.00)
	i) Picking, Filling	111.58 (24.35)	110.00 (19.48)	110.75 (36.36)	110.00 (42.04)
	ii) Depreciation of container (Tokri/Crate/Gunny bags)	151.28 (33.01)	231.00 (42.15)	73.50 (24.13)	48.00 (18.34)
	iii) Transportation cost	80.26 (17.51)	101.67 (18.01)	90.00 (29.54)	85.00 (32.48)
	iv) Loading/unloading charges	10.00 (2.18)	10.00 (1.77)	10.00 (3.28)	10.00 (3.82)
	v) Miscellaneous charges	3.00 (0.65)	2.33 (0.43)	1.50 (0.49)	8.67 (3.31)
	vi) Commission	71.28 (15.55)	0.00 (0.00)	0.00 (0.00)	0.00 (0.00)
2.	Marketing cost incurred by the wholesaler	**0.00** (0.00)	**60.00** (10.95)	**0.00** (0.00)	**0.00** (0.00)
	i) Loading/unloading charges	0.00 (0.00)	10.00 (1.82)	0.00 (0.00)	0.00 (0.00)
	ii) Transportation cost	0.00 (0.00)	50.00 (9.12)	0.00 (0.00)	0.00 (0.00)
3	Marketing cost incurred by the retailer	**30.91** (6.74)	**33.00** (5.84)	**18.88** (6.20)	**0.00** (0.00)
	i) Transportation cost	6.84 (1.49)	9.00 (1.59)	0.00 (0.00)	0.00 (0.00)
	ii) Loading/unloading charges	10.66 (2.33)	10.00 (1.77)	5.25 (1.72)	0.00 (0.00)
	iii) Shop/Rehri charges	3.41 (0.74)	4.00 (0.71)	3.63 (1.19)	0.00 (0.00)
	iv) Cost of plastic bags	10.00 (2.18)	10.00 (1.77)	10.00 (3.28)	0.00 (0.00)
	Total Marketing Cost (1+2+3)	**458.31**	**548.00**	**304.63**	**261.67**

Figure in parentheses are the percentage of total marketing cost of their respective channels.

channel-III (₹73.50) and channel-IV (₹48.00). The per quintal miscellaneous charges were found to be highest in channel IV (₹8.67) followed by channel I (₹3.00), channel II (₹2.33) and channel III (₹1.50).The commission agent charged an amount of ₹71.28 per quintal in channel I. The wholesaler incurred an expenditure of ₹60.00 per quintal in channel II in which transportation cost was higher (₹50.00) than loading/ unloading (₹10.00) of fruit. As far as the retailer was concerned, the cost incurred was ₹6.84, ₹10.66, ₹3.41 and ₹10.00 as transportation cost, loading/unloading charges,

shop/rehri charges and cost of plastic bags, respectively for channel I. Similarly for channel II, it was ₹9.00, ₹10.00, ₹4.00 and ₹10.00, respectively where as for channel III, it was ₹5.25, ₹3.63 and ₹10.00, respectively for loading/unloading charges, shop/ rehri charges and cost of plastic bags. Thus, the total marketing cost involved in marketing of citrus in the selected sample area was ₹458.31, ₹548.00, ₹304.63 and ₹261.67 for channel I, II, III and IV, respectively.

The channel-wise decomposition of marketing costs components for citrus fruit in Samba district is given in Table 4.36. The distribution channel-II (Producer → Wholesaler → Retailer → Consumer) was not found in sample area of Samba district. The marketing costs involved for the ultimate disposal of the produce like (picking, filling, cost of container, transportation cost and loading/unloading charges) was ₹438.65, ₹264.00 and ₹226.67 per quintal for channel I, III and IV, respectively in which ₹112.57, ₹111.00 and ₹111.67, respectively was for picking and filling, ₹10.00

Table 4.36: Channel-wise Decomposition of Marketing Costs Components for Citrus Fruit in Samba District

(₹/q)

Sl.No.	*Functionary*	*Channel–I*	*Channel–III*	*Channel–IV*
1	Marketing cost incurred by the producer	**438.65** (93.41)	**264.00** (93.15)	**226.67** (100.00)
	i) Picking, Filling	112.57 (23.97)	111.00 (39.17)	111.67 (49.27)
	ii) Depreciation of container (Tokri/Crate/Gunny bags)	194.00 (41.31)	84.00 (29.64)	48.17 (21.25)
	iii) Transportation cost	49.43 (10.53)	56.00 (19.76)	48.33 (21.32)
	iv) Loading/unloading charges	10.00 (2.13)	10.00 (3.53)	10.00 (4.41)
	v) Miscellaneous charges	3.43 (0.73)	3.00 (1.06)	8.50 (3.75)
	vi) Commission	69.22 (14.74)	0.00 (0.00)	0.00 (0.00)
2.	Marketing cost incurred by the retailer	**30.95** (6.59)	**19.40** (6.85)	**0.00** (0.00)
	i) Transportation cost	4.00 (0.85)	0.00 (0.00)	0.00 (0.00)
	ii) Loading/unloading charges	12.57 (2.68)	5.00 (1.76)	0.00 (0.00)
	iii) Shop/Rehri charges	4.38 (0.93)	4.40 (1.55)	0.00 (0.00)
	iv) Cost of plastic bags	10.00 (2.13)	10.00 (3.53)	0.00 (0.00)
	Total Marketing Cost (1+2)	**469.60**	**283.40**	**226.67**

Figure in parentheses are the percentage of total marketing cost of their respective channels.

per quintal in each channel for loading/unloading and ₹49.43, ₹56.00 and ₹48.33, respectively as transportation cost. The table further indicated that the grower also incurred expenditure on depreciation of container which was highest in case of channel I (₹194.00) followed by channel III (₹84.00) and channel IV (₹48.17). The per quintal miscellaneous charges were found to be highest in channel IV (₹8.50) followed by channel I (₹3.43) and channel III (₹3.00). Expenditure of ₹69.22 per quintal were incurred as commission to commission/forwarding agent in channel I. The retailer incurred about ₹4.00 towards transportation charges in channel I for carrying the

Table 4.37: Channel-wise Decomposition of Marketing Costs Components for Citrus Fruit in Jammu Region (Overall)

(₹/q)

Sl.No.	*Functionary*	*Channel–I*	*Channel–II*	*Channel–III*	*Channel–IV*
1.	Marketing cost incurred by the producer	**549.65** (94.61)	**476.60** (83.42)	**344.53** (94.71)	**274.29** (100.00)
	i) Picking, Filling etc	111.78 (21.88)	111.60 (19.53)	111.86 (30.75)	111.67 (40.71)
	ii) Depreciation of container (Tokri/Crate/Gunny bags)	198.15 (34.11)	194.25 (34.00)	139.70 (38.40)	58.89 (24.48)
	iii) Transportation cost	159.97 (27.53)	158.75 (27.79)	80.58 (22.15)	85.48 (21.47)
	iv) Loading/unloading charges	10.00 (1.96)	10.00 (1.75)	10.00 (2.75)	10.00 (3.65)
	v) Miscellaneous charges	3.00 (0.52)	2.00 (0.35)	2.39 (0.66)	8.25 (3.01)
	vi) Commission	66.75 (11.49)	0.00 (0.00)	0.00 (0.00)	0.00 (0.00)
2.	Marketing cost incurred by the wholesaler	**0.00** (0.00)	**63.13** (11.05)	**0.00** (0.00)	**0.00** (0.00)
	i) Loading/unloading charges	0.00 (0.00)	10.00 (1.75)	0.00 (0.00)	0.00 (0.00)
	ii) Transportation cost	0.00 (0.00)	53.13 (9.30)	0.00 (0.00)	0.00 (0.00)
3	Marketing cost incurred by the retailer	**31.33** (6.13)	**31.57** (5.53)	**19.25** (5.29)	**0.00** (0.00)
	i) Transportation cost	5.73 (1.12)	7.44 (1.30)	0.00 (0.00)	0.00 (0.00)
	ii) Loading/unloading charges	11.63 (2.28)	10.00 (1.75)	5.02 (1.38)	0.00 (0.00)
	iii) Shop/Rehri charges	3.97 (0.78)	4.13 (0.72)	4.23 (1.16)	0.00 (0.00)
	iv) Cost of plastic bags	10.00 (1.96)	10.00 (1.75)	10.00 (2.75)	0.00 (0.00)
Total Marketing Cost (1+2+3)		**580.98**	**571.30**	**363.78**	**274.29**

Figure in parentheses are the percentage of total marketing cost of their respective channels.

fruits to shop. The retailer also incurred a cost of ₹12.57 and ₹5.00 towards loading/ unloading and ₹4.38 and ₹4.40 towards shop/rehri charges for channel I and III, respectively while as the cost of plastic bags came out to be ₹10.00 per quintal in both the channels (I and III). The total marketing cost of the retailer came out to be ₹30.95 and ₹19.40, respectively in channel I and III. Thus, the total marketing cost involved in marketing of citrus in the selected sample area was ₹469.60, ₹283.40 and ₹226.67 for channel I, III and IV, respectively.

The channel-wise decomposition of marketing costs components for citrus fruit for Jammu region as a whole is given in Table 4.37. The table revealed that on an average, marketing expenses involved were ₹549.65, ₹476.60, ₹344.53 and ₹274.29 per quintal for the producer in channel I, II, III and IV, respectively, in which approximately ₹112.00 and ₹10.00 was worked out for picking and filling, loading/ unloading for each channel, ₹159.97, ₹158.75, ₹80.58 and ₹85.48, respectively as transportation cost and ₹198.15, ₹194.25, ₹139.70 and ₹58.89, respectively in case of depreciation of container. The per quintal miscellaneous charges were found to be

Selling of citrus by an orchardist

Figure 4.19a: An Overview of Marketing for Citrus Fruits

Selling of fruits through retailer at his shop/rehri

Figure 4.19b: An Overview of Marketing for Citrus Fruits

highest in channel IV (₹8.25) followed by channel I (₹3.00), channel III (₹2.39) and channel II (₹2.00). The table further revealed that the commission agent charged ₹66.75 per quintal in channel I. Marketing cost borne by the wholesaler in channel II was ₹63.13 per quintal in which transportation cost involved was the highest (₹53.13 per quintal) followed by loading/unloading (₹10.00 per quintal). At the retailers' level, the respective amounts of transportation cost, loading and unloading charges, shop/rehri charges and cost of plastic bags amounted to ₹5.73 and ₹7.44 for channel I and II, respectively, ₹11.63, ₹10.00 and ₹5.02 for I, II and III channels, ₹3.97, ₹4.13 and ₹4.23 for channel I, II and III, respectively and ₹10.00 in each channel, respectively. The total marketing costs incurred by the retailer was ₹31.33, ₹31.57 and ₹19.25 in channel I, II and III, respectively. In channel IV the producer had borne the whole marketing cost (₹274.29) as in this channel there was no intermediary and producer directly sold the produce to the consumer. Thus, the total marketing cost involved in marketing of citrus in the selected sample area was ₹580.98, ₹571.30, ₹363.78 and ₹274.29 for channel I, II, III and IV, respectively.

4.4.4.2.2. Marketing Loss of Citrus

In this study an attempt has been made to estimate marketing losses of citrus at different stages of marketing. But during the survey, it was found that marketing losses of the citrus fruit were negligible at producers' level. However, Table 4.38 revealed that the marketing losses ranged between ₹16.00 to ₹20.00 per q at wholesalers' level in channel II of Jammu, Rajouri and Kathua districts and accounted for 0.94 per cent, 0.83 per cent and 0.85 per cent of the consumers' price. The average marketing losses of all the channels of study area at retailers' level ranged between ₹16.50 to ₹20.50 per q.

Table 4.38: Decomposition of Marketing Loss Components for Citrus Fruit in Various Districts of Jammu Region

(₹/q)

Sl.No.	Functionary	Jammu	Rajouri	Kathua	Samba
1.	Average marketing loss at the producers' level	0.00	0.00	0.00	0.00
2.	Average marketing loss at the wholesalers' level (spoilage/wastage)	18.75	20.00	16.67	0.00
3.	Average marketing loss at the retailers' level (spoilage/wastage)	20.33	17.57	16.71	16.75
	Total marketing loss (1+2+3)	**39.08**	**37.57**	**33.38**	**16.75**

4.4.4.2.3. Price Spread and Marketing Margins

The price spread is the gap between the price paid by the consumer and the price received by the orchardist at a particular time because from the producer, it has to pass through various agencies before it reaches the final consumer. The distributional channel in addition to some charges for the services they perform, some margin money is also expected by them in that transaction. Therefore, it is worthwhile to examine as to what share of the rupee paid by the consumer is received by the producer.

The price spread as per cent of consumers' rupee for different market functionaries of citrus under different channels in Jammu district of Jammu region is presented in Table 4.39. The citrus growers of Jammu district received the net price of about ₹590.00/q, ₹691.00/q, ₹954.29/q and ₹996.00/q which were 28.92 per cent, 31.48 per cent, 48.20 per cent and 76.62 per cent of the price paid by the consumer for channel I, II, III and IV, respectively. The producers' sale price of citrus was ₹1350.00/q in channel I while as it was ₹1243.50/q, ₹1380.00/q and ₹1300.00/q in channel II, III and IV, respectively. The table further revealed that the per quintal marketing cost incurred by the orchardist was ₹760.00 in case of channel I followed by channel II (₹552.50), channel III (₹425.71) and channel IV (₹304.00), which accounted for 38.19 per cent in channel I, followed by channel II (26.37 per cent), channel III (21.50 per cent) and channel IV (15.59 per cent) of the consumer rupee.

The marketing costs borne by the wholesaler was 3.22 per cent of the consumer rupee and his margin was about 11.88 per cent. The marketing loss incurred by the wholesaler was ₹18.75/q (0.89 per cent). The wholesalers' sale price to retailer was

₹1578.64/q. Similarly, the retailers' sale price was ₹2040.00/q, ₹2195.00/q and ₹1980.00/q in channel I, II and III, respectively. In case of retailer, marketing cost on loading and unloading, transportation, shop/rehri charges and cost of plastic bags worked out to be 1.53 per cent in case of channel I followed by 1.42 per cent in case of channel II and 0.93 per cent in case of channel III. Margin of retailer was found to be maximum in case of channel-I (₹636.80/q *i.e.* 32.00 per cent) and in case of channel II and III, it was observed to be ₹565.60/q (26.99 per cent), ₹564.25/q (28.50 per cent).

Table 4.39: Price Spread of Citrus Fruit Under Different Marketing Channels in Jammu District

(₹/q)

Sl.No.	Particulars	Channel–I	Channel–II	Channel–III	Channel–IV
1.	Net price received by the producer	590.00	691.00	954.29	996.00
2.	Marketing cost incurred by the producer	760.00 (38.19)	552.50 (26.37)	425.71 (21.50)	304.00 (15.59)
3.	Producers' sale price	1350.00 (66.18)	1243.50 (56.65)	1380.00 (69.12)	1300.00 (100.00)
4.	Marketing cost incurred by the wholesaler	0.00 (0.00)	67.50 (3.22)	0.00 (0.00)	0.00 (0.00)
5.	Marketing loss incurred by the wholesaler	(0.00) (0.00)	18.75 (0.89)	(0.00) (0.00)	(0.00) (0.00)
6.	Margin of wholesaler	0.00 (0.00)	248.89 (11.88)	0.00 (0.00)	0.00 (0.00)
7.	Wholesalers' sale price/ retailers' purchase price	0.00 (0.00)	1578.64 (71.92)	0.00 (0.00)	0.00 (0.00)
8.	Marketing cost incurred by the retailer	30.49 (1.53)	29.77 (1.42)	18.50 (0.93)	0.00 (0.00)
9.	Marketing loss incurred by the Retailer	22.75 (1.14)	21.00 (1.00)	17.25 (0.87)	0.00 (0.00)
10.	Margin of retailer	636.80 (32.00)	565.60 (26.99)	564.25 (28.50)	0.00 (0.00)
11.	Retailers' sale price	2040.00 (100.00)	2195.00 (100.00)	1980.00 (100.00)	0.00 (0.00)
12.	Price paid by consumer	2040.00 (100.00)	2195.00 (100.00)	1980.00 (100.00)	1300.00 (100.00)
13.	Producers' share in consumers' rupee	0.29 (28.92)	0.31 (31.48)	0.48 (48.20)	0.77 (76.62)
	Total marketing margin	**636.80**	**814.49**	**564.25**	**0.00**

Figures in parentheses are the percentages of price paid by consumer.

The data given in Table 4.40 revealed the price spread of citrus under different channels in Rajouri district. The per quintal net price received by the citrus growers of Rajouri district was about ₹859.50, ₹900.00, ₹997.95 and ₹1026.00 which was about 46.46 per cent, 47.85 per cent, 54.53 per cent and 82.74 per cent of the price paid by the consumer for channel I, II, III and IV, respectively. The producer had to bear

Table 4.40: Price Spread of Citrus Fruit Under Different Marketing Channels in Rajouri District

(₹/q)

Sl.No.	Particulars	Channel–I	Channel–II	Channel–III	Channel–IV
1.	Net price received by the producer	859.50	900.00	997.95	1026.00
2.	Marketing cost incurred by the producer	396.50 (21.43)	250.00 (13.29)	242.05 (13.23)	214.00 (17.26)
3.	Producers' sale price	1256.00 (67.89)	1150.00 (61.15)	1240.00 (67.76)	1240.00 (100.00)
4.	Marketing cost incurred by the wholesaler	0.00 (0.00)	55.00 (2.92)	0.00 (0.00)	0.00 (0.00)
5.	Marketing loss incurred by the wholesaler	0.00 (0.00)	20.00 (1.06)	0.00 (0.00)	0.00 (0.00)
6.	Margin of wholesaler	(0.00) (0.00)	215.00 (11.43)	0.00 (0.00)	0.00 (0.00)
7.	Wholesalers' sale price/ retailers' purchase price	0.00 (0.00)	1440.00 (76.57)	0.00 (0.00)	0.00 (0.00)
8.	Marketing cost incurred by the retailer	33.28 (1.80)	34.50 (1.83)	19.50 (1.07)	0.00 (0.00)
9.	Marketing loss incurred by the Retailer	21.50 (1.16)	20.76 (1.10)	10.45 (0.57)	0.00 (0.00)
10.	Margin of retailer	540.00 (29.19)	415.00 (22.07)	560.00 (30.60)	0.00 (0.00)
11.	Retailer's sale price	1850.78 (100.00)	1910.25 (100.00)	1829.95 (100.00)	0.00 (0.00)
12.	Price paid by consumer	1850.00 (100.00)	1880.75 (100.00)	1829.95 (100.00)	1240.00 (100.00)
13.	Producer's share in consumer's rupee	0.46 (46.46)	0.48 (47.85)	0.54 (54.53)	0.83 (82.74)
Total marketing margin		**540.00**	**630.00**	**560.00**	**0.00**

Figures in parentheses are the percentages of price paid by consumer.

expenses to the extent of 21.43 per cent, 13.29 per cent, 13.23 per cent and 17.26 per cent out of the price paid by the consumer in channel I, II, III and IV, respectively. The producers' sale price was ₹1256.00/q in channel I as compared to ₹1150.00/q in channel II and ₹1240.00/q each in channel III and IV, respectively. A comparison of different channels showed that producers' share in the consumers' rupee was the highest in case of channel-IV (82.74 per cent) followed by channel-III (54.53 per cent), channel-II (47.85 per cent) and channel-I (46.86 per cent). The marketing costs borne by the wholesaler was 2.92 per cent (₹55.00/q) of the consumer rupee and his margin was about 11.43 per cent (₹215.00/q) when a grower directly sold to wholesaler. The marketing loss incurred by the wholesaler was ₹20.00/q (1.06 per cent). The wholesalers' sale price to retailer was ₹1440.00/q. Retailers' marketing cost worked out to be ₹33.28/q (1.80 per cent) in case of channel I, ₹34.50/q (1.83 per cent) in case

of channel II and ₹19.50/q (1.07 per cent) in case of channel III. Retailers' margin was 29.19 per cent, 22.07 per cent and 30.60 per cent in case of channel I, II and III, respectively. The consumer had to pay the highest price of ₹1880.75/q via channel II as compared to the ₹1850.00/q, ₹1829.95/q and ₹1240.00/q via channel I, III and IV, respectively.

The price spread for different market functionaries of citrus under different channels in Kathua district of Jammu region is shown in Table 4.41. The citrus growers received per quintal net price of about ₹998.30, ₹920.00, ₹1089.25 and ₹1113.33 which were about 48.70 per cent, 43.32 per cent, 53.39 per cent and 80.97 per cent of the price paid by the consumer for channel I, II, III and IV, respectively. The producers' sale price was ₹1425.70/q in channel I as compared to ₹1375.00/q each for other three channels *i.e.*, channel II, III and IV. The per quintal marketing costs incurred by the orchardist was ₹427.40 in case of channel I, ₹455.00 in channel II, ₹285.75 in channel

Table 4.41: Price Spread of Citrus Fruit Under Different Marketing Channels in Kathua District

(₹/q)

Sl.No.	Particulars	Channel–I	Channel–II	Channel–III	Channel–IV
1.	Net price received by the producer	998.30	920.00	1089.25	1113.33
2.	Marketing cost incurred by the producer	427.40 (20.85)	455.00 (21.43)	285.75 (14.01)	261.67 (19.03)
3.	Producers' sale price	1425.70 (69.55)	1375.00 (64.75)	1375.00 (67.40)	1375.00 (100.00)
4.	Marketing cost incurred by the wholesaler	0.00 (0.00)	60.00 (2.83)	0.00 (0.00)	0.00 (0.00)
5.	Marketing loss incurred by the wholesaler	0.00 (0.00)	16.67 (0.78)	0.00 (0.00)	0.00 (0.00)
6.	Margin of wholesaler	0.00 (0.00)	250.00 (11.77)	0.00 (0.00)	0.00 (0.00)
7.	Wholesalers' sale price/ retailer's purchase price	0.00 (0.00)	1701.67 (80.13)	0.00 (0.00)	0.00 (0.00)
8.	Marketing cost incurred by the retailer	30.91 (1.51)	33.00 (1.55)	18.88 (0.93)	0.00 (0.00)
9.	Marketing loss incurred by the Retailer	18.38 (0.90)	19.00 (0.89)	12.75 (0.63)	0.00 (0.00)
10.	Margin of retailer	575.00 (28.05)	420.00 (19.78)	633.37 (31.05)	0.00 (0.00)
11.	Retailers' sale price	2050.00 (100.00)	2123.67 (100.00)	2040.00 (100.00)	0.00 (0.00)
12.	Price paid by consumer	2050.00 (100.00)	2123.67 (100.00)	2040.00 (100.00)	1375.00 (100.00)
13.	Producers' share in consumers' rupee	0.49 (48.70)	0.43 (43.32)	0.53 (53.39)	0.81 (80.97)
	Total marketing margin	**575.00**	**670.00**	**633.37**	**0.00**

Figures in parentheses are the percentages of price paid by consumer.

III and ₹261.67 in channel IV. The percentage of the marketing cost out of the consumer rupee was highest in channel II (21.43 per cent) followed by channel I (20.85 per cent), channel IV (19.03 per cent) and channel III (14.01 per cent).

The table further revealed that the marketing margin of the wholesaler was about 11.77 per cent of the price paid by consumer out of which about 2.83 per cent of the consumers' rupee was paid as expenses on loading/unloading and transportation cost. The marketing loss incurred by the wholesaler was ₹16.67/q (0.78 per cent). The wholesalers' sale price to the retailer was ₹1701.67/q (80.13 per cent). Similarly, the retailers' sale price was ₹2050.00/q, ₹2123.67/q and ₹2040.00/q in channel I, II and III, respectively. The per quintal marketing costs borne by the retailer were found to be ₹30.91 (1.51 per cent) in case of channel I, ₹33.00 (1.55 per cent) in case of channel II and ₹18.88 (0.93 per cent) in case of channel III. The margin of the retailer was found to be 28.05 per cent, 19.78 per cent and 31.05 per cent of the consumers' price in case of channel I II and III, respectively.

Table 4.42: Price Spread of Citrus Fruit Under Different Marketing Channels in Samba District

(₹/q)

Sl.No.	*Particulars*	*Channel–I*	*Channel–III*	*Channel–IV*
1.	Net price received by the producer	945.90	1036.00	1073.33
2.	Marketing cost incurred by the producer	438.65 (20.40)	264.00 (13.07)	226.67 (17.44)
3.	Producers' sale price	1384.55 (64.40)	1300.00 (64.36)	1300.00 (100.00)
4.	Marketing cost incurred by the retailer	30.95 (1.44)	19.40 (0.96)	0.00 (0.00)
5.	Marketing loss incurred by the retailer	19.50 (0.91)	14.00 (0.69)	0.00 (0.00)
6.	Margin of retailer	715.00 (33.26)	686.60 (33.99)	0.00 (0.00)
7.	Retailers' sale price	2150.00 (100.00)	2020.00 (100.00)	0.00 (0.00)
8.	Price paid by consumer	2150.00 (100.00)	2020.00 (100.00)	1300.00 (100.00)
9.	Producers' share in consumers' rupee	0.44 (44.00)	0.51 (51.29)	0.83 (82.56)
	Total marketing margin	**715.00**	**686.60**	**0.00**

Figures in parentheses are the percentages of price paid by consumer.

The price spread for different market functionaries of citrus under different channels in Samba district of Jammu region is shown in Table 4.42. In the study area of this district, the wholesaler was not involved directly in any channel but sometimes wholesaler himself worked as forwarding/commission agent. The growers of Samba district received the net price of about ₹945.90/q, ₹1036.00/q and ₹1073.33/q which

was about 44.00, 51.29 and 82.56 per cent of the price paid by the consumer for channel I, III and IV, respectively. The producers' sale price was ₹1384.55/q in channel I and ₹1300.00/q each in channel III and channel IV. The marketing costs borne by the orchardist was ₹438.65/q in case of channel I followed by channel III (₹264.00/q) and channel IV (₹226.67/q). The marketing cost in the consumer rupee was highest in channel I (20.40 per cent) followed by channel IV (17.44 per cent) and channel III (13.07 per cent).

The retailers' sale price was ₹2150.00/q and ₹2020.00/q in channel I and III, respectively. The marketing costs borne by the retailer were found to be ₹30.95/q (1.44 per cent) in case of channel I and ₹19.40 (0.96 per cent) in case of channel III. Marketing margin of the retailer was found to be 33.26 per cent and 33.99 per cent of the consumers' price in case of channel I and III, respectively.

The price spread as per cent of consumers' rupee for different market functionaries of citrus under different channels in overall Jammu region is presented in Table 4.43. The average per quintal net price received by the citrus growers of Jammu region was about ₹855.35, ₹864.40, ₹995.47 and ₹1055.71 which was about 42.75 per cent, 39.33 per cent, 52.04 per cent and 79.38 per cent of the price paid by the consumer for channel I, II, III and IV, respectively. The producer had to bear marketing expenses to the extent of about ₹479.65/q (23.97 per cent), ₹475.60/q (21.64 per cent), ₹344.53/q (18.01 per cent) and ₹274.29/q (20.62 per cent) of consumer price in channel I, II, III and IV, respectively. The producers' sale price was ₹1335.00/q in channel I where as it was ₹1340.00/q both in channel II and III and ₹1330/q in channel IV, respectively.

The marketing costs borne by the wholesaler was 2.87 per cent (₹63.13/q) of the consumer rupee and when the produce was sold directly by the grower to the wholesaler (wholesalers') the margin realized was about 9.78 per cent (₹215.00/q). The marketing loss incurred by the wholesaler was ₹18.13/q (0.82 per cent). The wholesalers' sale price to the retailer was ₹1430.00/q. In case of retailer, marketing cost on loading/unloading, transportation, shop/rehri charges and cost of plastic bags were found to be ₹31.33/q (1.57 per cent) in case of channel I followed by ₹31.57/q (1.44 per cent) in case of channel II and ₹19.25/q (1.01 per cent) in case of channel III. The marketing loss incurred by the retailer was ₹20.53/q (1.03 per cent), ₹20.25/q (0.92 per cent) and ₹13.61/q (0.71 per cent in case of channel I, II and III, respectively. The margin of the retailer was about 30.69 per cent, 23.20 per cent and 28.23 per cent in case of channel I, II and III, respectively. The consumer of the study area had to pay the higher price of ₹2030.00/q via channel II as compared to ₹2000.86/q, ₹1912.86/q and ₹1330.00/q via channel I, III and IV, respectively.

4.4.4.2.4 Marketing Efficiency in Different Channels

The marketing efficiency is an important tool and therefore requires more attention. The marketing efficiency of different marketing channels of Jammu district is shown in Table 4.44. The citrus growers received highest net return per quintal from channel IV (₹996.00) followed by channel III (₹954.29), channel II (₹591.00) and channel I (₹540.00) whereas the marketing cost per quintal was found to be highest in channel I (₹790.49) followed by channel II (₹649.77), channel III (₹244.21) and channel

IV (₹304.00). Marketing loss was worked out to be ₹22.75 in channel I, ₹39.75 in channel II and ₹17.25 in channel III. Channel IV with marketing efficiency 3.28 found to be the most efficient marketing channel for the growers of citrus followed by channel III (1.16), channel II (0.39) and channel I (0.37).

Table 4.43: Price Spread of Citrus Fruit Under Different Marketing Channels in Jammu Region (overall)

(₹/q)

Sl.No.	Particulars	Channel–I	Channel–II	Channel–III	Channel–IV
1.	Net price received by the producer	855.35	864.40	995.47	1055.71
2.	Marketing cost incurred by the producer	479.65 (23.97)	475.60 (21.64)	344.53 (18.01)	274.29 (20.62)
3.	Producers' sale price	1335.00 (66.72)	1340.00 (60.96)	1340.00 (70.05)	1330.00 (100.00)
4.	Marketing cost incurred by the wholesaler	0.00 (0.00)	63.13 (2.87)	0.00 (0.00)	0.00 (0.00)
5.	Marketing loss incurred by the wholesaler	0.00 (0.00)	18.13 (0.82)	0.00 (0.00)	0.00 (0.00)
6.	Margin of wholesaler	0.00 (0.00)	215.00 (9.78)	0.00 (0.00)	0.00 (0.00)
7.	Wholesalers' sale price/ retailers' purchase price	0.00 (0.00)	1430.00 (65.06)	0.00 (0.00)	0.00 (0.00)
8.	Marketing cost incurred by the retailer	31.33 (1.57)	31.57 (1.44)	19.25 (1.01)	0.00 (0.00)
9.	Marketing loss incurred by the Retailer	20.53 (1.03)	20.25 (0.92)	13.61 (0.71)	0.00 (0.00)
10.	Margin of retailer	614.00 (30.69)	510.00 (23.20)	540.00 (28.23)	0.00 (0.00)
11.	Retailers' sale price	2000.86 (100.00)	2198.08 (100.00)	1912.86 (100.00)	0.00 (0.00)
12.	Price paid by consumer	2000.86 (100.00)	2030.00 (100.00)	1912.86 (100.00)	1330.00 (100.00)
13.	Producers' share in consumers' price	0.43 (42.75)	0.39 (39.33)	0.52 (52.04)	0.79 (79.38)
	Total marketing margin	**614.00**	**725.00**	**540.00**	**0.00**

Figures in parentheses are the percentages of price paid by consumer.

Table 4.45 revealed the efficiency of different marketing channels of Rajouri district. The citrus growers of Rajouri district received highest net return per quintal from channel IV (₹1026.00) followed by channel III (₹997.95), channel II (₹900.00) and channel IV (₹859.50) whereas the marketing cost per quintal was found to be highest in channel I (₹529.78) followed by channel II (₹319.50), channel III (₹264.73) and channel IV (₹230.00). Marketing loss was worked out to be ₹21.50 in channel I, ₹40.76 in channel II and ₹10.45 in channel III. The results indicated that channel IV

with marketing efficiency of 4.46 was most efficient followed by Channel III (1.19), channel II (0.91) and channel I (0.79).

Table 4.44: Marketing Efficiency of Different Channels for Citrus Fruit in Jammu District

Particulars	*Channel-I*	*Channel-II*	*Channel-III*	*Channel-IV*
Net price received by farmer(₹/q)	540.00	591.00	954.29	996.00
Marketing margin (₹/q)	636.80	814.49	564.25	0.00
Marketing cost (₹/q)	790.49	649.77	244.21	304.00
Marketing loss (₹/q)	22.75	39.75	17.25	0.00
Marketing efficiency	**0.37**	**0.39**	**1.16**	**3.28**

Table 4.45: Marketing Efficiency of Different Channels for Citrus Fruit in Rajouri District

Particulars	*Channel-I*	*Channel-II*	*Channel-III*	*Channel-IV*
Net price received by farmer(₹/q)	859.50	900.00	997.95	1026.00
Marketing margin (₹/q)	540.00	630.00	560.00	0.00
Marketing cost (₹/q)	529.78	319.50	264.73	230.00
Marketing loss (₹/q)	21.50	40.76	10.45	0.00
Marketing efficiency	**0.79**	**0.91**	**1.19**	**4.46**

Table 4.46: Marketing Efficiency of Different Channels for Citrus Fruit in Kathua District

Particulars	*Channel-I*	*Channel-II*	*Channel-III*	*Channel-IV*
Net price received by farmer(₹/q)	998.30	920.00	1089.25	1113.33
Marketing margin (₹/q)	575.00	620.00	633.37	0.00
Marketing cost (₹/q)	458.31	548.00	297.63	296.67
Marketing loss (₹/q)	18.38	35.67	12.75	0.00
Marketing efficiency	**0.95**	**0.76**	**1.15**	**3.75**

The marketing efficiency indices of different marketing channels of Kathua district are shown in Table 4.46. The citrus growers received highest net return per quintal from channel IV (₹1113.33) followed by channel III (₹1089.25), channel I (₹998.30) and channel II (₹920.00) whereas the marketing cost per quintal was found to be highest in channel II (₹548.00) followed by channel I (₹458.31), channel III (₹297.63) and channel IV (₹296.67). Marketing loss was worked out to be ₹18.38 in channel I, ₹35.67 in channel II and ₹12.75 in channel III. Highest marketing efficiency was noticed in Channel IV (3.75) followed by channel III (1.15), channel I (0.95) and channel II (0.76).

Table 4.47 revealed the efficiency of different marketing channels of Samba district. The citrus growers received highest net return per quintal from channel IV (₹1073.33) followed by channel III (₹1036.00), and channel I (945.90) whereas the marketing cost per quintal was found to be highest in channel I (₹469.60) followed by channel III (₹283.40) and channel IV (₹226.67). Marketing loss was worked out to be ₹19.50 in channel I and ₹14.00 in channel III. The marketing efficiency was highest in channel IV (4.74) followed by channel III (1.05) and channel I (0.79).

Table 4.47: Marketing Efficiency of Different Channels for Citrus Fruit in Samba District

Particulars	*Channel-I*	*Channel-III*	*Channel-IV*
Net price received by farmer(₹/q)	945.90	1036.00	1073.33
Marketing margin (₹/q)	715.00	686.60	0.00
Marketing cost (₹/q)	469.60	283.40	226.67
Marketing loss (₹/q)	19.50	14.00	0.00
Marketing efficiency	**0.79**	**1.05**	**4.74**

The marketing efficiency indices of different marketing channels of overall Jammu region are shown in Table 4.48. The citrus growers of the overall Jammu region received highest net return per quintal from channel IV (₹1055.71) followed by channel III (₹995.47), channel II (₹864.40) and channel I (₹855.35) whereas the marketing cost per quintal was found to be highest in channel II (₹571.30) followed by channel I (₹510.98), channel III (₹363.78) and channel IV (₹274.29). Marketing loss per quintal was worked out to be ₹20.53 in channel I, ₹38.38 in channel II and ₹13.61 in channel III. Channel IV had the marketing efficiency of 3.85 followed by channel III (1.09), channel I (0.75) and channel II (0.65).

Table 4.48: Marketing Efficiency of Different Channels for Citrus Fruit in Jammu Region (Overall)

Particulars	*Channel-I*	*Channel-II*	*Channel-III*	*Channel-IV*
Net price received by farmer(₹/q)	855.35	864.40	995.47	1055.71
Marketing margin (₹/q)	614.00	725.00	540.00	0.00
Marketing cost (₹/q)	510.98	571.30	363.78	274.29
Marketing loss (₹/q)	20.53	38.38	13.61	0.00
Marketing efficiency	**0.75**	**0.65**	**1.09**	**3.85**

4.4.5. Price Behaviour

4.4.5.1. Trends in Arrivals and Prices of Citrus in Narwal Market of Jammu District

Seasonal variation in the arrivals and prices of citrus *i.e.* orange, kinnow and lemon in the secondary and terminal markets is a well known phenomenon. Climatic conditions as well as human and institutional factors influence the market arrivals.

Very low bargaining power of the cultivators, lack of credit and storage facilities, pressing demand for ready cash and ignorance of the market information force the cultivators to sell their produce immediately after harvest. It was observed that the produce brought to the market after the harvest was treated as of better quality because of low keeping quality and farmers receive better price.

In addition to making an indepth study of growers' production process and marketing of citrus fruits, it was thought better to have an idea of some regulated market of Jammu region so that one could find the amount of citrus fruit supplied by the growers of Jammu region and also imported from other states of India. In Jammu region, there is not a single regulated market. Though Narwal market is not the regulated market but is a well organized market. Therefore, Narwal market was selected to have an idea about the supply and prices of citrus (orange, kinnow and lemon). The average monthly arrivals and prices for the period of last 5 years *i.e.* 2005-06 to 2009-10 are shown below in the different tables separately.

Table 4.49: Average Monthly Arrivals and Prices of Orange, Kinnow and Lemon in Narwal Market of Jammu (2005-06 to 2009-10)

Particulars	*Fruit-wise Arrivals (qs.)*			*Fruit-wise Prices (₹/q.)*		
	Orange	*Kinnow*	*Lemon*	*Orange*	*Kinnow*	*Lemon*
April	4916.60	-	1833.33	3002.50	-	2562.33
May	2383.33	-	3788.80	3779.00	-	2274.40
June	2735.00	-	7293.00	3400.00	-	1707.75
July	-	-	4430.40	-	-	1515.20
August	-	-	2980.20	-	-	1929.60
September	3778.67	-	2115.80	2441.67	-	2498.40
October	2593.40	-	2601.00	2323.50	-	2312.00
November	4093.00	3010.25	1759.00	2435.50	1422.00	1710.00
December	8895.40	10116.00	1695.80	2266.75	1308.60	1413.80
January	7005.80	6671.00	1947.00	2499.75	1487.00	1897.60
February	12339.40	7679.00	6460.00	2601.75	1604.20	1960.20
March	5930.40	2545.00	2328.00	2832.75	1829.20	2565.60

Notes: The months with no arrivals were excluded from the calculation.

Simple average was used to find out monthly arrivals and prices.

Source: Directorate of Horticulture, Planning and Marketing, Narwal, Jammu.

The Table 4.49 indicated the average monthly arrivals and prices of orange, kinnow and lemon in Narwal market of Jammu for the period 2005-06 to 2009-10. It depicted that the highest market arrivals of orange in Narwal market were recorded during the month of February (12339.40 qtls) and lowest in May (2383.33 qtls). However, the main season for arrivals of local orange in Narwal market was from December to February. The prices move contrary to arrivals of being the lowest in December (₹2266.75/q) and the highest in the month of May (₹3779.00/q). The table

also depicted that there was neither local orange supply nor import of orange during July and August months.

The table further revealed that the highest market arrivals of kinnow were recorded during the month of December (10116.00 qtls) and lowest in March (2545.00 qtls). The prices were recorded lowest in December (₹1308.60/q) and the highest in the month of March (₹1829.20/q).

The average monthly arrivals and prices of Lemon in Narwal market for the same period are also presented in the sane table and revealed that the highest market arrivals of lemon were recorded during the month of June (7293.00 qtls) and lowest in December (1695.00 qtls). The prices moved contrary to arrivals of being the lowest in December (₹1413.80/q) and the highest in the month of September (₹2498.40/q). The average monthly arrivals and prices of orange, kinnow and lemon are also shown graphically in Figure 4.20.

Table 4.50: Seasonal Indices of Arrivals and Prices of Orange, Kinnow and Lemon in Narwal Market of Jammu (2005-06 to 2009-10)

Particulars	*Fruit-wise Seasonal Index of Arrivals*			*Fruit-wise Seasonal Index of Price*		
	Orange	*Kinnow*	*Lemon*	*Orange*	*Kinnow*	*Lemon*
April	89.93	-	56.08	108.85	-	126.29
May	43.59	-	115.89	137.00	-	112.10
June	50.03	-	223.07	123.26	-	84.17
July	-	-	135.51	-	-	74.68
August	-	-	91.16	-	-	95.11
September	69.12	-	64.72	88.52	-	123.14
October	47.44	-	79.56	84.24	-	113.95
November	74.87	50.14	53.80	88.30	92.93	84.28
December	162.71	168.48	51.87	82.18	85.52	69.68
January	128.14	111.10	59.55	90.63	97.18	93.53
February	225.70	127.89	197.59	94.32	104.84	96.61
March	108.47	42.39	71.21	102.70	119.54	126.45

Note: The months with no arrivals were excluded from the calculation.

4.4.5.2. Seasonal Fluctuation in Citrus Arrivals and Prices

Table 4.50 depicted the seasonal indices of arrivals and prices of orange, kinnow and lemon in Narwal market of Jammu and revealed that indices of arrivals of orange in Narwal market were recorded maximum during the month of February (225.70) and lowest in May (43.59). The arrivals started picking up after September. The prices moved contrary to arrivals, being the lowest in the December (82.18) and the highest in May (137.0). It has been noticed that the prices started recovering from the month of January and reached the highest point during the month of May. The table further revealed that Narwal market did not witness the arrival of kinnow during April to

October. The arrivals of kinnow started picking up after November and were the highest during the month of December (168.48) and lowest in March (42.39). The prices moved contrary to arrivals, being the lowest in the December (85.52) and the highest in March (119.54). The situation in case of lemon was slightly different as compared to orange and kinnow. The arrivals of lemon in Narwal market were found to be maximum in the month of June (223.07) and lowest in December (51.87). Consequently, the prices were at their lowest ebb in the month of December. It has been noticed that the prices started recovering from the month of January and reached the highest point during the month of April (126.29). It was also found that the prices of lemon slumped due to higher volume of market arrivals during the month of June (84.17) and July (74.68). The seasonal indices for orange, kinnow and lemon are also shown graphically in Figure 4.21.

4.4.6. Constraints Faced by the Orchardists in Production and Marketing of Citrus

An attempt was also made to understand the problems that were faced by the growers in production and marketing of citrus in the area under study.

4.4.6.1 Production Constraints

The numbers of production and marketing constraints faced by citrus growers in Jammu, Rajouri, Kathua and Samba district of Jammu region and also of Jammu region as a whole are presented in Table 4.51 and Figure 4.22 and 4.23. The table indicated that at production level, 91.67 per cent of the respondents in the Jammu district expressed the problem of lack of finance and credit facilities followed by non availability of quality seedlings (79.17 per cent), inadequate irrigation facilities (75.00 per cent), non availability of well decomposed farmyard manure (70.83 per cent) and lack of latest technical knowledge (58.33 per cent). Also 43.75 per cent of the respondents were not using the recommended quantity of pesticides because of their high cost. In addition to this there were 41.67 per cent of the orchardists who were facing the problem of non availability of labour during peak period while 33.33 per cent opined that labour cost was high. The table further indicated that production problems like occurrence of diseases, educated members go outside and did not take interest in farming and lack of latest technical knowledge also accounted for 41.67 per cent, 29.17 per cent and 58.33 per cent, respectively. Whereas in Rajouri district, inadequate irrigation facilities (95.83 per cent), lack of latest technical knowledge (93.75 per cent), lack of finance and credit facilities (89.58 per cent), non availability of quality seedlings (87.50 per cent) and farmyard manure (83.33 per cent) were major problems which were faced by the respondents at production level. Also 37.50 per cent of the respondents were not using the recommended quantity of pesticides because of their high cost. In addition to this problems like non availability of labour during peak period, high labour cost, occurrence of diseases and educated members go outside varied from 20 per cent to 42 per cent of the respondents.

The table further indicated that in Kathua district the highest number of respondents (83.33 per cent) expressed the problem of educated members go outside and did not take interest in farming. Lack of finance and credit facilities, non availability of quality seedlings, high labour cost, inadequate irrigation facilities and

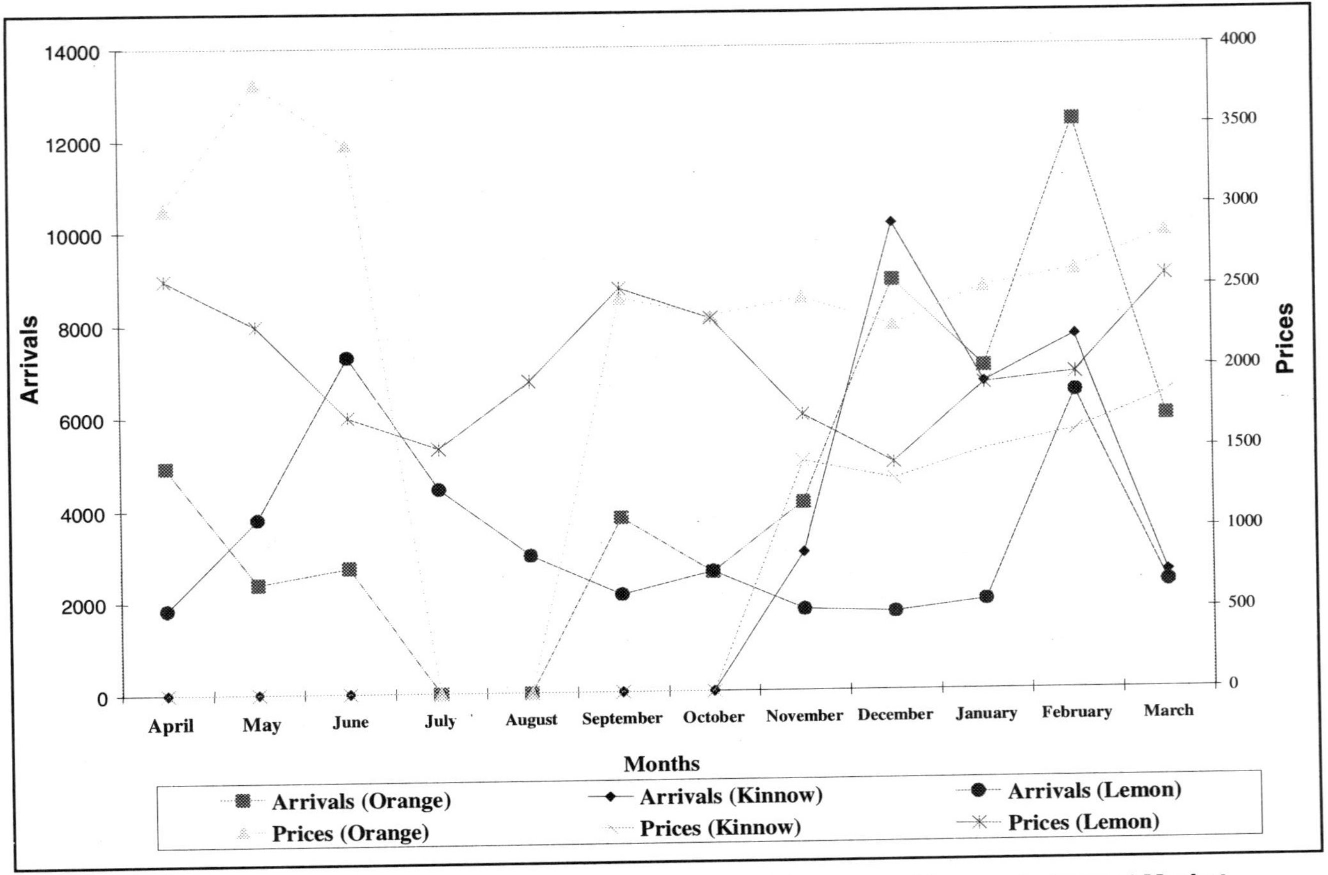

Figure 4.20: Trends in Arrivals and Prices of Orange, Kinnow and Lemon in Narwal Market (2005-06 to 2009-10)

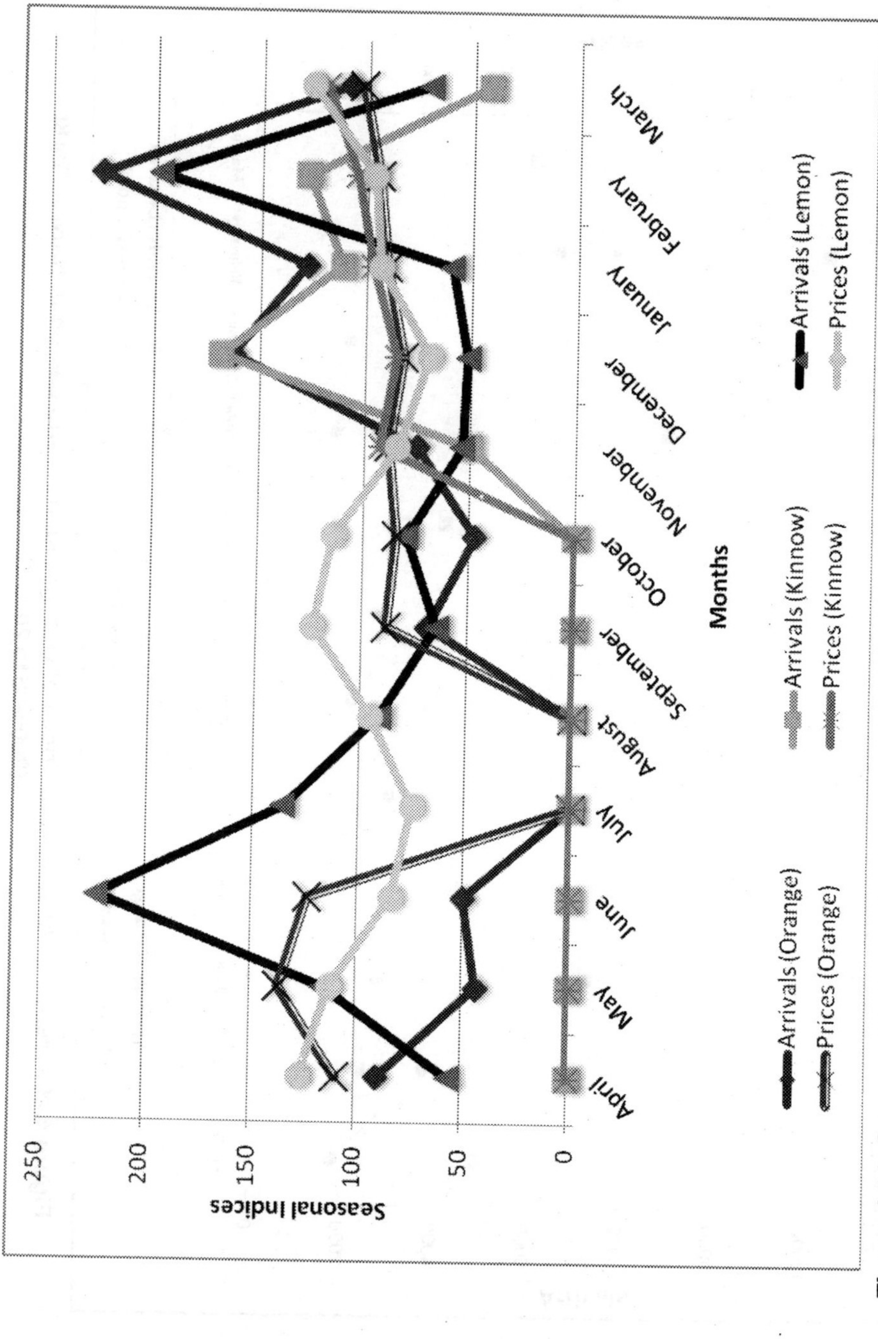

Figure 4.21: Seasonal Indices of Arrivals and Prices of Orange, Kinnow and Lemon in Narwal Market (2005-06 to 2009-10)

non availability of FYM were other major problems which accounted for 70.83 per cent, 66.67 per cent, 54.17 per cent, 52.08 and 50.00 per cent, respectively. In addition to this there were 39.58 per cent of the orchardists who were facing the problem of lack of latest technical knowledge while 31.25 per cent opined non availability of labour during peak period as major constraint. In case of Samba district, it was observed that 79.17 per cent of the respondents expressed the problem of lack of finance and credit facilities. Educated members go outside and did not take interest in farming, non availability of quality seedlings and FYM were other major problems which accounted for 77.08 per cent, 62.50 per cent and 60.42 per cent, respectively. In addition to this there were 52.08 per cent of the orchardists who were facing the problem of non availability of labour during peak period while 39.58 per cent opined that irrigation facilities were inadequate. The table further indicated that production problems like high labour cost, lack of latest technical knowledge, high cost of pesticides and occurrence of diseases also accounted for 35.42 per cent, 33.33 per cent, 29.17 per cent and 8.33 per cent, respectively.

In general, the numbers of production constraints faced by citrus growers in all the districts of Jammu region as a whole revealed that 82.81 per cent of the respondents in the sample area expressed the problem of lack of finance and credit facilities. Non availability of good quality seedlings and farmyard manure, inadequate irrigation facilities and lack of latest technical knowledge were other major problems which accounted for 73.96 per cent, 66.15 per cent, 65.63 per cent and 56.25 per cent, respectively. Also 52.60 per cent of the respondents were facing the problem that the educated members of their family go outside and did not take interest in farming. Besides there were 41.15 per cent of the orchardists who were facing the problem of high labour cost while 37.50 per cent opined that labour was short during peak period. The table further indicated that production problems like high cost of pesticides and occurrence of diseases also accounted for 33.33 per cent and 20.83 per cent, respectively.

4.4.6.2. Marketing Constraints

At marketing level, lack of processing units and marketing societies was the major constraint as expressed by 100.00 per cent respondents of the Jammu district. Moreover, lack of market information, un-organised marketing and low price paid to farmers and not getting remunerative price for the produce were also a major constraints expressed by 93.75 per cent, 87.50 per cent and 83.33 per cent, respectively. About 79.17 per cent of the respondents complained about less demand of fruits because of competition of other fruits and 33.33 per cent pointed out the problem of high perishability of fruits. The problem of cheating in marketing by the middlemen in the form of malpractices, high and undue marketing margins and deductions in the market were found to be 37.50 per cent. Costly packing material was reported by 33.33 per cent of the total sample orchardists as a major constraint followed by high cost of transportation (31.25 per cent), non availability of market (33.33 per cent), high commission charges (29.17 per cent) and packages not returned to the growers (8.33 per cent). Only 4.17 per cent growers opined that, there exists a problem of non receipt of payment in time. Whereas in Rajouri district lack of processing units and marketing societies was the major constraint as expressed by 95.83 per cent

respondents followed by lack of market information (85.42 per cent), non availability of market (66.67 per cent) and not getting remunerative price for the produce (62.50 per cent). Un-organised marketing and low price paid to farmers and costly packing material were also major constraints expressed by 60.42 per cent and 56.25 per cent, respectively. About 50.00 per cent of the respondents complained about less demand of fruits because of competition of other fruits and 41.67 per cent pointed out the problem of high perishability of fruits. The problems of cheating in marketing by the middlemen were found to be 45.83 per cent. High commission charges (41.67), high cost of transportation (37.50 per cent) and packages not returned to the growers (12.50 per cent) were reported by sample orchardists as major constraints. Only two per cent orchardists opined that there exists a problem of non receipt of payment in time.

The table also revealed that lack of processing units and marketing societies were the major constraints in Kathua district as expressed by 100.00 per cent respondents. Moreover, not getting remunerative price for the produce and cheating by middlemen were also major constraints expressed by 70.83 per cent and 68.75 per cent, respectively. About 60.42 per cent of the respondents complained about high cost of transportation and 52.08 per cent pointed out the problem of lack of market information. The problem of costly packing material and unorganised marketing and low price paid to farmers were found to be 47.92 per cent and 70.83 per cent,

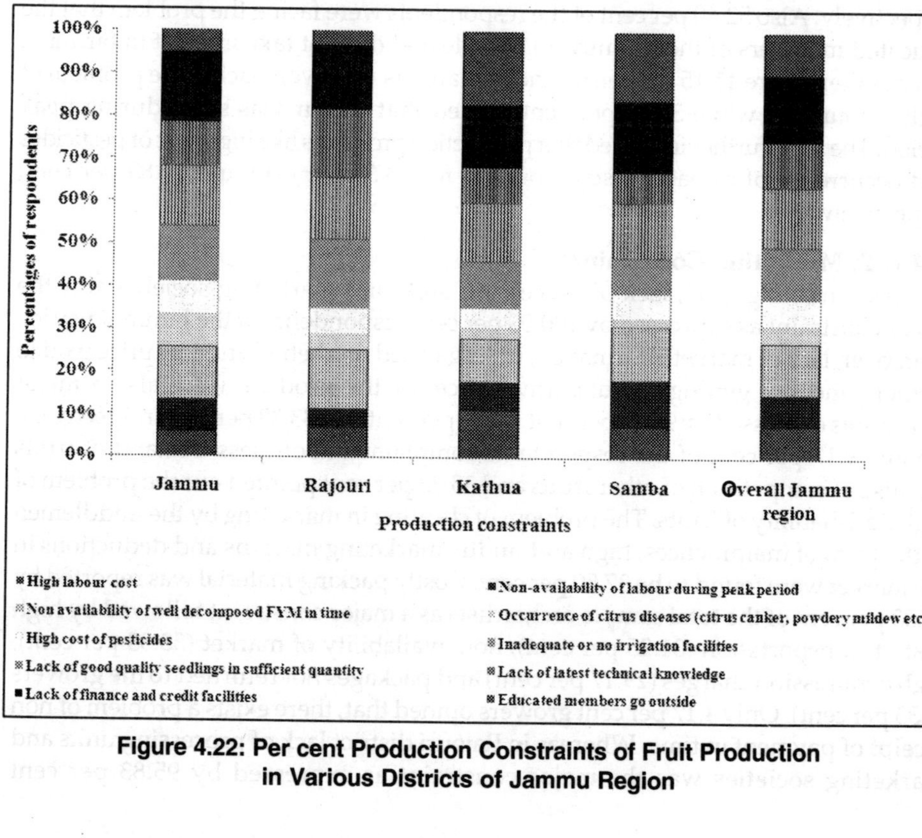

Figure 4.22: Per cent Production Constraints of Fruit Production in Various Districts of Jammu Region

Table 4.51: Constraints Faced by the Sample Orchardists in Production and Marketing of Citrus in Jammu, Rajouri, Kathua and Samba District of Jammu Region

Sl.No.	Constraints	Number of Respondents (N)				
		Jammu (N=48)	Rajouri (N=48)	Kathua (N=48)	Samba (N=48)	Overall Jammu Region (N=192)
	A. Production Problems					
1	High labour cost	16 (33.33)	20 (41.67)	26 (54.17)	17 (35.42)	79 (41.15)
2.	Non-availability of labour during peak period	20 (41.67)	12 (25.00)	15 (31.25)	25 (52.08)	72 (37.50)
3.	Non availability of well decomposed FYM in time	34 (70.83)	40 (83.33)	24 (50.00)	29 (60.42)	127 (66.15)
4.	Occurrence of citrus diseases (citrus canker, powdery mildew etc)	20 (41.67)	10 (20.83)	6 (12.50)	4 (8.33)	40 (20.83)
5.	High cost of pesticides	21 (43.75)	18 (37.50)	11 (22.92)	14 (29.17)	64 (33.33)
6.	Inadequate or no irrigation facilities	36 (75.00)	46 (95.83)	25 (52.08)	19 (39.58)	126 (65.63)
7.	Lack of good quality seedlings in sufficient quantity	38 (79.17)	42 (87.50)	32 (66.67)	30 (62.50)	142 (73.96)
8.	Lack of latest technical knowledge	28 (58.33)	45 (93.75)	19 (39.58)	16 (33.33)	108 (56.25)
9.	Lack of finance and credit facilities	44 (91.67)	43 (89.58)	34 (70.83)	38 (79.17)	159 (82.81)
10.	Educated members go outside	14 (29.17)	10 (20.83)	40 (83.33)	37 (77.08)	101 (52.60)

Contd...

Table 4.51–*Contd...*

Sl.No.	*Constraints*	*Number of Respondents (N)*				
		Jammu (N=48)	*Rajouri (N=48)*	*Kathua (N=48)*	*Samba (N=48)*	*Overall Jammu Region (N=192)*
	B. Marketing Problems					
1.	Not getting remunerative price for the produce	40 (83.33)	30 (62.50)	34 (70.83)	40 (83.33)	144 (75.00)
2.	Packing material is costly	16 (33.33)	27 (56.25)	23 (47.92)	27 (56.25)	93 (48.44)
3.	Packages are not returned to the growers	4 (8.33)	6 (12.50)	0 (0.00)	2 (4.17)	12 (6.25)
4.	Less demand of fruits because of competition of other fruits	38 (79.17)	24 (50.00)	18 (37.50)	25 (52.08)	105 (54.69)
5.	Cheating by middlemen	18 (37.50)	22 (45.83)	33 (68.75)	30 (62.50)	103 (53.65)
6.	High cost of transportation	15 (31.25)	18 (37.50)	29 (60.42)	35 (72.92)	97 (50.52)
7.	High commission charges	14 (29.17)	20 (41.67)	17 (35.42)	15 (31.25)	66 (34.38)
8.	Non receipt of payment in time	2 (4.17)	1 (2.08)	0 (0.00)	0 (0.00)	3 (1.56)
9.	Lack of market information	45 (93.75)	41 (85.42)	25 (52.08)	10 (20.83)	121 (63.02)
10.	Un-organised marketing and low price paid to farmers	42 (87.50)	29 (60.42)	34 (70.83)	34 (70.83)	125 (65.10)
11.	High perishability of the fruits	16 (33.33)	20 (41.67)	20 (41.67)	18 (37.50)	74 (38.54)
12.	Non-availability of market	16 (33.33)	32 (66.67)	0 (0.00)	1 (2.08)	49 (25.52)
13.	Lack of processing units and co-operative societies	48 (100.00)	46 (95.83)	48 (100.00)	44 (91.67)	186 (96.88)

Figures in parentheses are the percentages of the total number of respondents of the respective district.

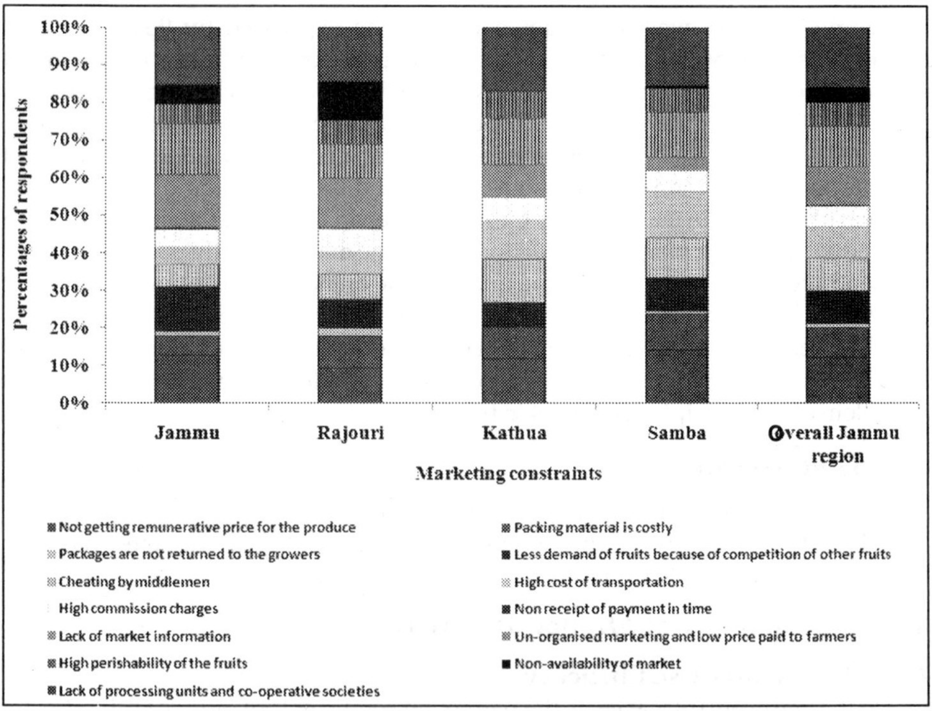

Figure 4.23: Per cent Marketing Constraints of Fruits Grown in Various Districts of Jammu Region

respectively. High commission charges were reported by 35.42 per cent of the total sample orchardists as a marketing constraint. The constraints like packages not returned to the growers, non receipt of payment in time and non-availability of market were not present in the Kathua district as per the opinion of the sample orchardists. In sample area of Samba district at marketing level, lack of processing units and marketing societies was the major constraint as expressed by 91.67 per cent respondents. Moreover, not getting remunerative price for the produce and un-organised marketing and low price paid to farmers were also a major constraints expressed by 83.33 per cent and 70.83 per cent, respectively. About 62.50 per cent of the respondents complained about cheating in marketing by the middlemen in the form of malpractices, high and undue marketing margins and deductions in the market and 56.25 per cent pointed out the problem of costly packing material. The problem of less demand of fruits because of competition of other fruits was found to be 52.08 per cent. High cost of transportation was reported by 72.92 per cent of the total sample orchardists as a major constraint followed by high commission charges (31.25 per cent), lack of market information (20.83 per cent) and packages not returned to the growers (4.17 per cent). Only 2.08 per cent growers opined that, there exists a problem of non availability of market.

In general, the numbers of marketing constraints faced by citrus growers in all the districts of Jammu region as a whole indicated that, lack of processing units and

marketing societies was the major constraint as expressed by 96.88 per cent respondents. Moreover, not getting remunerative price for the produce, un-organised marketing and low price paid to farmers and lack of market information were also major constraints expressed by 75.00 per cent 65.10 per cent and 63.02 per cent, respectively. About 54.69 per cent of the respondents complained about less demand of fruits because of competition of other fruits and 53.65 per cent pointed out the problem of cheating in marketing by the middlemen in the form of malpractices, high and undue marketing margins and deductions in the market. Costly packing material was reported by 48.44 per cent of the total sample orchardists as a major constraint followed by high cost of transportation (41.67 per cent), high commission charges (34.38 per cent), non availability of market (25.52 per cent), and packages not returned to the growers (6.25 per cent). Only 1.56 per cent respondents opined that, there exists a problem of non receipt of payment in time.

4.5. Discussion

The results obtained during the investigation entitled, "Economics of Production and Marketing of Citrus in Jammu region of Jammu and Kashmir state" has been discussed in this chapter under the following headings:

4.5.1. Economics of Orange Production

4.5.1.1. Resource Use Efficiency

In the age group of 5–9 years the regression coefficients of orange (Table 4.3) for human labour and manures + fertilizers with positive sign indicated that with one per cent increase in the use of these two inputs keeping all the other inputs constant, could increase the output of the crop to 0.77 per cent in case of human labour and 0.01 per cent in case of manures + fertilizers. The contribution of irrigation and plant protection were negative and non-significant. The regression coefficient of training/pruning was negative and significant. All the regression coefficients were less than unity thereby indicating diminishing returns. It could be seen from the Table 4.3 that marginal value productivity of human labour and manures + fertilizers were positive while that of irrigation, plant protection and training/pruning were negative hence indicating their excess use and should be avoided to check the fall of returns in the orange orchards in the age of 5 – 9 years. These results are in close conformity with Chinappa and Ramanna (1997).

In the age group of 10 – 14 years, the regression coefficients of orange (Table 4.4) for human labour, plant protection and training/pruning with positive sign indicated that with one per cent increase in the use of these inputs, keeping all the other inputs constant, could increase the output of the crop by 0.87 per cent in case of human labour, 0.05 per cent in case of plant protection and 0.02 per cent in case of training/pruning. The contribution of manures + fertilizers was however negative and significant thereby indicating that one per cent increase of this input, keeping all other inputs constant, could decrease the output by 0.06 per cent. It could be seen from the Table 4.4 that marginal value productivity of plant protection, training/pruning and human labour were positive thereby indicating that an increased used of these inputs could increase the output because they were sub optimally used while

that of manures + fertilizers was negative hence indicating its excess use and should be avoided to check the fall of returns in the orange orchards in the age of 10 – 14 years. These results are supported by Ahmad and Mustafa (2006).

In the age group of 15-19 years the regression coefficients of orange (Table 4.5) for human labour and manures + fertilizers with positive sign indicated that with one per cent increase in the use of these two inputs keeping all the other inputs constant, could increase the output of the orange crop to 0.87 per cent and 0.01 per cent, respectively, though non significance of manures + fertilizers revealed that this input had negligible role in the production of orange in this period. These results are supported by Utomakili and Molua (1998). The contribution of plant protection and training/pruning were negative and non-significant. These findings are in close conformity with Koujalagi *et al.* (1999). All the regression coefficients taken together were less than unity thereby indicating operation of diminishing returns. It could be seen from the Table 4.5 that marginal value productivity of human labour and manures + fertilizers were positive while that of plant protection and training/ pruning were negative hence indicating their excess use and should be avoided to check the fall of returns in the orange orchards in the age of 15 –19 years.

The results of regression coefficients of orange in the age group of 20 – 24 years (Table 4.6) showed that the regression coefficients of human labour and manures + fertilizers with positive sign indicated that with one per cent increase in the use of these two inputs, the output could be increased by 0.89 per cent and 0.05 per cent, respectively (because they were underutilized). The regression coefficient of plant protection was negative and non-significant. The marginal value productivity of manures + fertilizers and human labour indicated that an additional one rupee spent on them could add to the gross return by ₹59.54 per acre and ₹0.01 per acre, hence there was scope of investing more on manures + fertilizers and human labour. Similar findings were those of Hanumantharaya *et al.* (2009).

The results of regression coefficients of orange in the age group of 25 – 28 years (Table 4.7) showed that the regression coefficients of human labour and manures + fertilizers with positive significant sign indicated that with one per cent increase in the use of these two inputs, could have increased output by 0.72 per cent and 0.17 per cent, respectively. The regression coefficient of plant protection was negative and non significant. It could be observed that MVP of manures + fertilizers (203.46) and human labour indicated that an additional one rupee spent on manures + fertilizers could add to the gross return by ₹203.46, while that of human labour indicated that an additional one rupee spent on human labour could increase the returns by ₹0.37 per acre and hence there was scope of investing more on manures + fertilizers as well as on human labour.

The results of regression coefficients of overall group of orange (Table 4.8) depicted that regression coefficients of human labour, (manures + fertilizers) and plant protection with positive sign indicated that with one per cent increase in the use of these inputs, the output could be increased by 0.96 per cent in case of human labour and 0.01 per cent each in other two inputs. The regression coefficient of training/ pruning was negative but significant indicating that this input was significant in the

production of orange but it was used at more than optimum level, whereas that of irrigation was negative and non significant. The MVP as shown in Table 4.8 indicated that an additional one rupee spent on manures + fertilizers, human labour and plant protection could add to gross returns by ₹110.45, ₹0.19 and ₹0.08, respectively, hence there was scope of investing more on these inputs. The MVP of irrigation and training/ pruning indicated that with an additional one rupee invested on these inputs will reduce the gross returns and hence should be checked. These findings are in close conformity with Chinappa and Ramanna (1997).

4.5.1.2. Costs and Returns

The results of Table 4.9 *i.e.* first year per acre establishment costs of orange indicated that in case of orange orchards the cost incurred on digging, filling and planting (₹1405.79) was highest among the working cost as well as among the total costs whereas earned value of rented land (EVRL – ₹1334.63) was highest among the fixed costs. The first year establishment costs of orange orchards were ₹5089.08. In all the size groups studied, the cost on the digging, filling and planting was highest whereas that of plant protection (₹7.26) was lowest in case of working capital. The first year establishment costs were found to be highest in case of large orchards followed by medium, small and marginal orchards thereby indicating that first year establishment costs increased with the increase in size of orchards. It was observed due to more expenditure incurred on preparation of land and digging, filling and planting by large size orchardists. Similar findings were those of Gangwar and Singh (1998).

The results of year wise per acre establishment costs of orange (Table 4.10) indicated that high costs were required during the first year of establishment in all the holding sizes and in the successive years, the cost involvement ranged between 45-50 per cent of the first year costs till the plant started bearing. The first year establishment costs were found to be highest in case of large orchards followed by medium, small and marginal orchards. The costs in the second year and onwards were low mainly because only maintenance costs were required once the orchard was established. These findings are supported by Gangwar and Singh (1998).

The results of item wise and concept-wise per acre operational costs of orange (Table 4.11) indicated that overall cost A, cost B and cost C were ₹1060.88, ₹2700.44 and ₹3849.35, respectively, and all the three costs increased with the increase in the farm size from marginal to medium orchards except in large orchards and it was due to family human labour which was an important factor in determining the costs incurred on the maintenance of the orange orchards. The share of the family human labour (₹1148.91) was the highest among the working costs in all the size groups of orange orchards followed by hired human labour (₹563.09). Among the fixed costs, the costs on EVRL were the highest. Similar findings were those of Subrahmanyam (1987) and Gangwar *et al.* (1998).

The returns per year per acre (Table 4.12) increased as the age of the orange plant increased. The large orchardists had the highest returns among all the size of holdings for all the years under consideration. Therefore, larger the size of an orchard, more the profits which indicated the economics of large scale production. These results are in conformity with that of Gangwar *et al.* (2005).

4.5.1.3. Economic Viability

The results of economic viability of orange orchards (Table 4.13) revealed that the per acre net present value of orange orchards ranged from ₹3347.62 in marginal orchards to ₹5984.83 in small orchards, hence indicating that the net present value was highest in small orchards and lowest in marginal orchards. The internal rate of return ranging from 12.56 per cent in marginal orchards to 18.25 per cent in large orchards indicated that orange growing was a profitable enterprise. The benefit cost ratio calculated at cost C ranged from 2.01 in marginal orchards to 2.33 in large orchards, thus indicating that the medium and large orchardists got maximum returns for each rupee invested. The pay-back period of 7.2 to 8.1 indicated that the orchardists could have their investment back within a period of 7 years 2 months to 8 years 1 month. Similar findings were those of Gupta and George (1974).

4.5.2. Economics of Kinnow Production

4.5.2.1. Resource Use Efficiency

In the age group of 5 – 9 years, the regression coefficients of kinnow (Table 4.14) for human labour and manures + fertilizers with positively significant coefficient indicated that with one per cent increase in the use of these two inputs keeping all the other inputs constant, could increase the output by 0.32 per cent in both cases and at same time, both these inputs had significant contribution in the kinnow production. The regression coefficient of plant protection was however negative and non significant. It could be seen from the Table 4.14 that marginal value productivity of human labour, manures + fertilizers, irrigation and training/pruning were positive and showed that additional one rupee spent on these inputs, could add to gross returns by ₹0.10, ₹0.47, ₹2.05 and ₹0.06, respectively and hence there still existed a scope to invest more on these inputs. Similar findings were those of Iqbal (2009).

In the age group of 10 – 14 years, the regression coefficients of kinnow (Table 4.15) for human labour, manures + fertilizers and training/pruning with significantly positive sign indicated that with one per cent increase in the use of these inputs, keeping all the other inputs constant, could increase the output of the crop by 0.25 per cent, 0.57 per cent and 0.12 per cent, respectively. The contribution of plant protection was negative and non significant. Similar findings were those of Koujalagi *et al.* (1999). It could be further seen from the Table 4.15 that marginal value productivity of human labour, manures + fertilizers, and training/pruning with positive sign revealed that one rupee spent on these inputs could increase the returns by ₹0.003, ₹0.99 and ₹0.01, respectively while as plant protection with negative sign indicated its excess use.

In the age group of 15 – 19 years, the regression coefficients of kinnow (Table 4.16) for human labour, manures + fertilizers and plant protection with positive sign indicated their under utilization and with one per cent increase in the use of these inputs keeping all the other inputs constant, could increase the output of the kinnow crop to 0.23 per cent, 0.26 per cent and 0.14 per cent, respectively. The contribution of training/pruning was however positive but non-significant. All the regression coefficients were less than unity thereby indicating operation of diminishing returns. Here, the marginal value productivity of all the explanatory variables were positive

with their value at 0.01, 9.07, 0.02 and 0.07 in case of human labour, manures + fertilizers, plant protection and training/pruning, respectively which meant that there is still a scope in investing in these inputs. These finding are in close conformity with Iqbal (2009).

The results of regression coefficients of kinnow in the age group of 20 – 24 years (Table 4.17) showed that the regression coefficients of human labour, manures + fertilizers and plant protection with positive sign indicated that with one per cent increase in the use of these inputs, the output could be increased by 0.01 per cent, 0.71 per cent and 0.16 per cent, respectively. The marginal value productivity of all the explanatory variables was positive. This showed that additional one rupee spent on these inputs would add to the gross returns by ₹0.004, ₹0.01 and ₹0.02 in case of manures + fertilizers, plant protection and human labour, respectively. Ahmad and Mustafa (2006) also recorded the similar observations.

The results of regression coefficients of kinnow in the age group of 25–28 years (Table 4.18) showed that the regression coefficients of human labour and manures + fertilizers with positive sign indicated that with one per cent increase in the use of these two inputs, the return of output could be increased by 0.75 per cent and 0.03 per cent, respectively, though the human labour contribution was significant in the kinnow production while as manures + fertilizers contribution was not so significant in its production. The regression coefficient of plant protection was negative and non significant. The MVP of human labour (0.50) and manures + fertilizers (0.04) indicated that an additional one rupee spent on human labour could add to the gross return by ₹0.50, while that of manures + fertilizers indicated that an additional one rupee spent on it could increase the returns by ₹0.04 per acre and hence there was scope of investing more on human labour as well as on manures + fertilizers, but plant protection could decrease the returns by ₹1.08 per acre.

The results of regression coefficients of overall group of kinnow (Table 4.19) depicted that regression coefficients of human labour, irrigation, plant protection and training/pruning with positive sign indicated that with one per cent increase in the use of these inputs, the output could be increased by 0.03 per cent, 0.02 per cent, 0.02 per cent and 0.14 per cent, respectively though only human labour and training/pruning contributed significantly in kinnow production for the whole life of this orchard. The manures + fertilizers with negative and non significant regression coefficient implied its negligible contribution in the production of kinnow. The function analysis also revealed that 73.6 per cent of the total variation in gross returns was explained by the explanatory variables. These findings are in close conformity with Koujalagi *et al.* (1999). The MVP as shown in Table 4.19 indicated that an additional one rupee spent on human labour, plant protection, training/pruning and irrigation could add by ₹0.03, ₹0.01, ₹0.18 and ₹0.03 per acre, respectively, to gross returns, hence there was scope of investing more on these inputs.

4.5.2.2. Costs and Returns

The results of Table 4.20 *i.e.* first year per acre establishment costs of kinnow indicated that in case of kinnow orchards the cost incurred on digging, filling and planting (₹1620.69) was highest among working cost as well as among the total costs

whereas earned value of rented land (EVRL- ₹1105.35 was highest among the fixed costs. The first year establishment costs of kinnow orchards were ₹5298.32. In all the size groups studied, the cost on the digging, filling and planting was highest whereas that of plant protection (₹3.83) was lowest in case of working capital. The first year establishment costs were found to be highest in case of medium orchards followed by small and marginal orchards thereby indicating that first year establishment costs increases with the increase in the size of holding from marginal to medium orchards which was mainly due to more application of fertilizers and plant protection chemicals during planting by medium size orchardists. These results are in conformity with the Gangwar and Singh (1998) and Gangwar *et al.* (2005).

The results of year wise per acre establishment costs of kinnow (Table 4.21) indicated that high costs were required during the first year of establishment in all the size holdings and in the successive years 40-50 per cent of the first year costs (₹5298.32) were required till the plant started bearing. The first year establishment costs were found to be highest in case of medium orchards (₹5523.92) followed by small (₹5391.00) and marginal (₹5262.50) orchards. The costs in the second year and onwards were low mainly because the costs were required only for the aftercare. These findings are supported by Radha *et al.* (2000) and Gangwar (2005).

The results of item wise and concept wise operational costs of kinnow (Table 4.22) indicated that overall per acre cost A, cost B and cost C were ₹1142.29, ₹2550.61 and ₹3762.89, respectively, thereby indicating that all the three costs increased with the increase in the farm size. The results also concluded that family human labour was an important factor in determining the costs incurred on the maintenance of the kinnow orchards which worked out to be 30.46 per cent of the total cost incurred for the whole period. The per acre share of the family human labour (₹1212.28) was the highest among the working costs in all the size groups of kinnow orchards followed by hired human labour (₹732.31). Among the fixed costs, the costs on EVRL (₹1105.35) were the highest. The total operational costs increased with the increase in the size of orchards. These findings are in close conformity with Singh and Sayeed (2008) and Yeware *et al.* (2010).

The results of returns per year from kinnow orchards (Table 4.23) revealed that the returns per acre increased as the age of the plant increased. The returns per acre were highest in case of medium orchards (₹7965.50) followed by small (₹7125.35) and marginal orchards (₹6954.32) upto the age of 15 years and above. The overall returns per acre returns (₹7385.38) were also highest in case of medium orchards followed by small (₹6774.05) and marginal orchards (₹6562.11). The overall returns were ₹6632.07 per acre. Therefore in general, indicated that the larger the size of an orchard, more will be the returns which implies economies of large scale production. Thakur *et al.* (1986) and Sudha *et al.* (1988) reported similar findings.

4.5.2.3. Economic Viability

The results of economic viability of kinnow orchards (Table 4.24) revealed that the net present value ranged from ₹7467.53 per acre in marginal orchards to ₹11649.23 per acre in medium orchards, hence indicating that the net present value was highest in medium orchards and lowest in marginal orchards. The results are in conformity

with Gupta and George (1974). The internal rate of return ranging from 14.75 per cent in marginal orchards to 16.00 per cent in medium orchards indicated that kinnow growing was a profitable enterprise and the average rate of return per year for the whole period of the orchard will be 14.75 per cent for marginal orchard while as it will be 15.50 per cent for small and 16 per cent for medium orchards. The benefit cost ratio calculated at cost C ranged from 1.07 in small orchards to 1.65 in marginal and medium orchards, thus indicating that the marginal and medium orchardists got ₹1.65 for each rupee they invested. The pay-back period of 7.2 to 7.8 indicated that the orchardists could have their investment back within a period of 7 years 2 months to 7 years 8 months. These findings are in close conformity with Gangwar and Singh (1998).

4.5.3. Economics of Lemon Production

4.5.3.1. Resource Use Efficiency

In the age group of 5 – 9 years the regression coefficients of lemon (Table 4.25) for human labour, plant protection and training/pruning with positive sign indicated that with one per cent increase in the use of these three inputs keeping all the other inputs constant, could increase the output of the crop to 1.30 per cent, 0.01 per cent and 0.02 per cent, respectively, though plant protection and training/pruning had negligible contribution in production of lemon. The regression coefficient of manures + fertilizers was however negative (-0.85) and non significant. The irrigation coefficient though significant meaning that in this age group irrigation played an important role but it was overutilized and with one per cent increase in this input could reduce the output by 0.06 per cent. It could be seen from the Table 4.25 that marginal value productivity of human labour, plant protection and training/pruning were positive and showed that additional one rupee spent on these inputs, could add to gross returns by ₹0.10, ₹0.02 and ₹0.04, respectively and hence there was still a scope to invest more on these inputs. There was no need to invest more on irrigation and manures + fertilizers.

In the age group of 10 – 14 years the regression coefficients of lemon (Table 4.26) for human labour with significantly positive sign indicated that with one per cent increase in the use of this input, keeping all the other inputs constant could increase the output of the crop by 1.54 per cent. The contribution of manures + fertilizers, plant protection and training/pruning were however negative thereby indicating that one per cent increase of these input, keeping all other inputs constant, could decrease the output by 0.10 per cent in case of manures + fertilizers, 0.03 per cent in case of plant protection and 0.01 per cent in case of training/pruning. The marginal value productivity of human labour was positive while that of manures + fertilizers, plant protection and training/pruning were negative hence indicating their excess use and should be avoided to check the fall of returns in the lemon orchards in the age of 10 – 14 years because they could reduce the output by ₹0.05, ₹0.01 and ₹0.05, respectively with one additional rupee spent on them. These findings are in close conformity with Iqbal (2009).

In the age group of 15 – 19 years the regression coefficients of lemon (Table 4.27) for human labour and manures + fertilizers with positive sign indicated that with

one per cent increase in the use of these two inputs keeping all the other inputs constant, could increase the output of the lemon crop by 0.05 per cent and 0.97 per cent, respectively though in this age group, manures + fertilizers had significant contribution in lemon output while as human labour had negligible contribution. The contribution of plant protection was however significantly negative thereby indicating that one per cent increase of this input, keeping all other inputs constant, could decrease the output by 0.08 per cent. All the regression coefficients were less than unity thereby indicating operation of diminishing returns. It could be seen from the Table 4.27 that marginal value productivity of human labour and manures + fertilizers were positive while that of plant protection was negative, thereby indicating that with one additional one rupee spent on these inputs could increase the output by ₹0.07 and ₹0.08, respectively while as in case of plant protection output could reduce by ₹0.01.

The results of regression coefficients of overall group of lemon (Table 4.28) depicted that regression coefficients of human labour and manures + fertilizers with significantly positive sign indicated that with one per cent increase in the use of these inputs, keeping the other inputs constant, could increase the output by 0.45 per cent and 1.26 per cent, respectively. Similar findings were those of Khushk *et al.* The regression coefficient of irrigation, plant protection and training/pruning were negative and non significant. The MVP as shown in Table 4.28 indicated that an additional one rupee spent on human labour and manures + fertilizers could add to gross returns by ₹0.11 and ₹0.88, respectively, hence there was scope of investing more on these inputs. The MVP of irrigation, plant protection and training/pruning indicated that with an additional one rupee invested on them would have reduced the gross returns by ₹0.02, ₹59.71 and ₹0.04 and hence should be checked. These findings are in close conformity with Koujalagi *et al.* (1999).

4.5.3.2. Costs and Returns

The results of Table 4.29 *i.e.* first year per acre establishment costs of lemon indicated that in case of lemon orchards the cost incurred on digging, filling and planting (₹1028.06) was highest among working cost whereas earned value of rented land (EVRL – ₹1050.82) was highest among the fixed costs as well as among the total costs. The first year establishment costs of lemon orchards were ₹3821.59. The minimum expenditure was found to be on irrigation (₹67.78) as all the irrigation done in lemon in the study area was by canal irrigation by all the orchardists. As the number of orchardists in case of lemon were less, so it was not possible to study it under different size of holdings.

The results of year wise establishment costs of lemon (Table 4.30) indicated that high costs were required during the first year of establishment and in the successive years 48-52 per cent of the first year costs were required till the plant started bearing. The first year establishment costs were found to be ₹3821.59. The costs in the second year and onwards were low mainly because the costs were required only for the aftercare.

The results of item wise and concept-wise operational costs of lemon (Table 4.29) indicated that overall per acre cost A, cost B and cost C were ₹816.77, ₹2241.64

and ₹2930.27, respectively thereby indicating that family human labour was an important factor in determining the costs incurred on the maintenance of the lemon orchards. The share of the family human labour (₹688.63) was the highest among the working costs followed by hired human labour (₹597.16). Among the fixed costs, the costs on EVRL were the highest *i.e.*, ₹1050.82. These findings are in conformity with Singh and Sayeed (2008).

The results of returns per year from lemon orchards (Table 4.30) revealed that the returns per acre increased as the age of the plant increased. The returns upto 10 years were on an increase. The 11th – 15th year was peak maturity period of lemon orchard. After that the decreasing trend of the orchard started. The overall returns were ₹10475.36 which were found to be more as compared to orange and kinnow. These results were in close conformity with the results of Subrahmanyam (1986) and Sudha *et al.* (1988).

4.5.3.3. Economic Viability

The results of economic viability of lemon orchards (Table 4.31) revealed that the net present value was ₹5475.61 and the internal rate of return 20.80 per cent in overall lemon orchards, hence indicated that lemon growing was a profitable enterprise. The benefit cost ratio calculated at cost C was equal to 2.70 in overall lemon orchards, thus indicating that the lemon orchardists could get good returns for each rupee they invested. These findings were in close conformity with Supe *et al.* (2009). The pay-back period of 6.4 indicated that the orchardists could have their investment back within a period of 6 years and 4 months. The results are in conformity with Subrahmanyam (1986).

4.5.4. Marketing of Citrus

4.5.4.1. Marketing Channels in Selected Study Area

The results of Table 4.32 revealed the quantity of citrus sold through the different marketing channels in various districts of Jammu region. The total quantity sold through channel I, II, III and IV in Jammu region was worked out to be 2180.49, 662.29, 1264.59 and 374.81 quintals, respectively, which accounted for 48.65 per cent, 14.78 per cent, 28.21 per cent and 8.36 per cent, respectively. This shows that growers mainly used to dispose of their produce through commission/forwarding agent. Based on quantities of citrus marketed through different channels, it was noticed that the highest quantities were sold through channel I (Producer → Forwarding/Commission agent → Retailer → Consumer), thereby indicating prominence of channel I in the study area. The lowest quantities were sold through the channel IV (Producer → Consumer).

4.5.4.2. Marketing Costs, Loss, Margins and Price Spread

4.5.4.2.1. Marketing Cost of Citrus

The results of Table 4.33 revealed the marketing cost components of market functionaries such as producers, commission agents, wholesalers and retailers of Jammu district. The total marketing costs of selected channels were ₹790.49 per quintal for channel I, ₹649.77 for channel II, ₹444.21 for channel III and ₹304.00 for channel

IV which indicated that more the market functionaries or intermediaries, more will be the marketing cost involved. The reason for high marketing cost in channel I and channel II as compared to other channels was due to high transportation cost and depreciation of container. The high cost of picking and filling which ranged between 14.20 per cent of the total marketing cost in channel I to 36.51 per cent in channel IV was also one of the main reasons of high marketing cost incurred by the producer. The commission (8.54 per cent) was observed in case of channel I only. Wholesaler was present in channel II only and transportation cost which was found to be 85.19 per cent was the main factor of marketing cost and was incurred by wholesaler. It was further found that the cost incurred by retailer was maximum in plastic bags in channel I, II and III whereas in channel IV, the orchardists sold directly to the consumers and none of the cost were repeated *i.e.*, the marketing cost in case of channel IV was the lowest.

The results of Table 4.34 revealed the information regarding the marketing cost incurred by the producers, commission agents, wholesalers and retailers of Rajouri district. The total marketing costs incurred by these functionaries was highest in channel I (₹429.78 per quintal) followed by channel II (₹339.50 per quintal), channel III (₹261.55 per quintal) and channel IV (₹214.00 per quintal) implying that more the intermediaries in the market, more will be the marketing cost involved. Secondly, the marketing cost was more in Jammu district as compared to Rajouri district which was due to the high transportation cost and depreciation of container incurred by the producers' of Jammu district. The main reason for high marketing cost in channel I as compared to other channels lied in high transportation cost, depreciation of container and commission charged by the agents. The high cost of picking and filling in all the channels was also one of the main reasons of high marketing cost incurred by the producer (₹110.00 for each channel). Another component of total marketing cost was the cost incurred by wholesaler on transportation in channel-II only.

The results of Table 4.35 revealed the information regarding the marketing cost incurred by the producers, commission agents, wholesalers and retailers in Kathua district of Jammu region. The total marketing costs incurred by these functionaries was highest in channel II (₹548.00 per quintal) followed by channel I (₹458.31 per quintal), channel III (₹304.63 per quintal) and channel IV (₹261.67 per quintal). Although the practice of commission charged by commission agent was present in channel I only but the marketing cost incurred by the producer as well as total marketing cost was high in channel II and the high transportation cost and high cost incurred on depreciation of container was owing to non availability of forwarding/ commission agents in remote villages of Basohli block of Kathua district and they covered a long distance to sold their produce to wholesalers. The high cost of picking and filling which ranged between 19.48 per cent of the total marketing cost in channel II to 42.04 per cent in channel IV was also one of the main reasons of high marketing cost incurred by the producer. The wholesaler was present in channel II only and the marketing cost incurred by him on transportation was ₹50.00 per quintal. It was further found from the table that transportation, loading/unloading, cost of plastic bags were the main items of marketing cost incurred by the retailer in channel I, II and III except transportation cost component in channel II. In channel IV producer sold

directly to consumers and none of the cost were repeated *i.e.*, the marketing cost in case of channel IV was the lowest.

The results of Table 4.36 revealed the channel-wise information regarding the marketing cost incurred by the producers, commission agents and retailers in Samba district. The total marketing costs incurred by these functionaries was highest in channel I (₹469.60 per quintal) followed by channel III (₹283.40 per quintal), and channel IV (₹226.67 per quintal). The channel II was not present in the study area of the Samba district. The main reason for high marketing cost in channel I (₹469.60) as compared to channel III (₹283.40) and channel IV (226.67) lied in depreciation of container (41.31 per cent) and commission (14.74 per cent) charged by the agents in channel I as compared to 29.64 per cent and 21.25 per cent for depreciation of container in channel III and channel IV, respectively. The high cost of picking and filling which was worked out to be ₹112.57, ₹111.00 and ₹111.67 in channel I, channel III and channel IV, respectively, was also one of the main reasons of high marketing cost incurred by the producer. It was further found that the cost incurred by retailer was highest in case of plastic bags in channel I and III (₹10.00 each) whereas in channel IV, the orchardists sold directly to the consumers and none of the cost were repeated *i.e.*, the marketing cost in case of channel IV was the lowest.

The results of Table 4.37 revealed the marketing cost components of various market functionaries such as producers, commission agents, wholesalers and retailers in Jammu region (overall). The total marketing costs incurred by these functionaries was highest in channel I (₹580.98 per quintal) followed by channel II (₹571.30 per quintal), channel III (₹363.78 per quintal) and channel IV (₹274.29 per quintal). The miscellaneous charges were found to be highest in channel IV (₹8.25 per quintal) which was due to the cost incurred by producer on plastic bags and rehri charges. The marketing cost incurred by the producer as well as total marketing cost was high in channel I and the main reason for that was high transportation cost, high cost incurred on depreciation of container and commission charged by commission agents at producers' level. The high cost of picking and filling in all the channels was also one of the main reasons of high marketing cost incurred by the producer. The wholesaler was present in channel II only and the marketing cost incurred by him on transportation was ₹53.13 per quintal. It was further found from the table that transportation, loading/unloading, shop/rehri charges and cost of plastic bags were the main items of marketing cost incurred by the retailer in channel I, II and III except transportation cost component in channel II. In channel IV producer sold directly to consumers and none of the cost were repeated *i.e.*, the marketing cost in case of channel IV was the lowest. These findings are in close conformity with Lepeha *et al.* (1993).

4.5.4.2.2. Marketing Loss of Citrus

Losses in agriculture are observed at production stage, harvesting and marketing/ distribution stage. Marketing losses vary depending on climate, kind of fruits and the distance from the market and the point of production. The losses increased cumulatively as the produce moves from harvesting to its consumption level. The improper harvesting, handling, transportation and distribution result in their significant losses.

In general, substantial quantity of production was subjected to losses at various stages of marketing in the study area. The quantum of loss was influenced by several factors like faulty method of handling, packing (crude packing method instead of bamboo basket and jute bags), transportation and lack of knowledge for proper storage.

Thus, the study revealed that in the study area, the producers' were mostly free from the losses in terms of economic value during marketing due to short distant markets and less quantity available with them for marketing. Only wholesalers' and retailers' had to bear marketing losses that too for end spoilage and wastage only.

From the results of the Table 4.38, the marketing losses at wholesalers' and retailers' level were identified at different stages like loading/unloading and transportation. Commission/forwarding agents never take the title of produce so there were no losses at commission agent level. Losses at producers' level were so less in the study area that they were considered negligible. It was due to less perishability of the fruit and sale of the fruit in the nearby market which reduces the loss during transportation. The Table revealed that the marketing losses at wholesalers' level were ₹18.75 per q, ₹20.00 per q and ₹16.67 per q in Jammu, Rajouri and Kathua district, respectively. The average marketing losses of all the channels of study area at retailers' level were ₹20.33 per q, ₹17.57 per q, ₹16.71 per q, ₹16.75 per q in Jammu, Rajouri, Kathua and Samba districts, respectively. The more than 75 per cent contribution towards the total marketing losses was due to end spoilage and wastage at wholesalers' and retailers' level.

4.5.4.2.3. Price Spread and Marketing Margins

The results of the Table 4.39 for the price spread of citrus for the different channels of Jammu district indicated that producers' share in consumers' rupee was highest in channel IV (76.62 per cent) followed by channel III (48.20 per cent), channel II (31.48 per cent) and channel I (28.92 per cent) which revealed that direct sale in the local market resulted in higher share in the consumers' rupee where in both producers' and consumers' gained because net price received by the producers was highest in channel IV and the price paid by the consumer was lowest in this channel. The findings are in close conformity with Gill *et al.* (1980). The margin of the retailer was highest in channel I (32.00 per cent) followed by channel III (28.50 per cent) and channel II (26.99 per cent) which indicated that more the market functionaries, more will be the margin of retailer. The table also revealed that as the number of intermediaries goes on decreasing the retailers' sale price also decreased and was lowest in channel III. These results are supported by Auhurkar and Deole (1985) and Mondal (1986).

The results of the Table 4.40 for the price spread of citrus for the different channels of Rajouri district indicated that producers' share in consumers' rupee was highest in channel IV (82.74 per cent) followed by channel III (54.53 per cent), channel II (47.85 per cent) and channel I (46.46 per cent) which revealed that direct sale in the local market resulted in higher share in the consumers' rupee. These findings are supported by Mondal (1986). The margin of the retailer was highest in channel III (30.60 per cent) followed by channel I (29.19 per cent) and channel II (22.07 per cent). But the total marketing margin was highest in channel II due to the presence of one

more intermediary *i.e.*, wholesaler. The channel IV was found to be better both for producer and consumer point of view with highest net price received by the producer and lowest price paid by consumer. These results are supported by Shah *et al.* (2010).

The results of the Table 4.41 for the price spread of citrus for the different channels of Kathua district indicated that producers' share in consumers' rupee was highest in channel IV (80.97 per cent) followed by channel III (53.39 per cent), channel I (48.70 per cent) and channel II (43.32 per cent) which revealed that direct sale in the local market resulted in higher share for the producer in the consumers' rupee. The results are in close conformity with the results of Mondal (1986). The wholesaler was present in channel II only and retained the margin of 11.77 per cent of the price paid by consumer. The margin of the retailer was highest in channel III (31.05 per cent) followed by channel I (28.05 per cent) and channel II (19.78 per cent). These results are supported by Ladaniya *et al.* (2003). But the total marketing margin was highest in channel II due to the presence of one more intermediary *i.e.*, wholesaler. The channel IV was found to be better both for producer and consumer point of view with highest net price received by the producer (₹1113.33 per quintal) and lowest price paid by consumer (₹1375.00 per quintal). Thus it becomes pertinent to have direct sale of the commodities and try to reduce the number of market intermediaries.

The results of the Table 4.42 for the price spread of citrus for the different channels of Samba district indicated that channel II was absent in the study area of the Samba district. The table revealed that producers' share in consumers' rupee was highest in channel IV (82.56 per cent) followed by channel III (51.29 per cent) and channel I (44.00 per cent) which revealed that direct sale in the local market resulted in highest share of the producer in the consumers' rupee. The wholesaler as such was not present but sometimes commission agent acted as wholesaler. The margin of the retailer was highest in channel I (₹715.00 per quintal) followed by channel III (₹686.60 per quintal). The best channel both for producer and consumer point of view was found to be channel IV in which producer got maximum share (82.56 per cent) of consumers' rupee and consumer purchased the fruit at minimum price (₹1300.00 per quintal).

The results of the Table 4.43 for the price spread of citrus for the different marketing channels of Jammu region (overall) indicated that producers' share in consumers' rupee was highest in channel IV (79.38 per cent) followed by channel III (52.04 per cent), channel I (42.75 per cent) and channel II (39.33 per cent) which revealed that direct sale resulted in highest share of producer in the consumers' rupee. The results are in close conformity with Mondal (1986) and also supported by Tomar *et al.* (1997). Thus lesser the number of market intermediaries, more the profit the producer will reap which is also clear from the net price received by the producer *i.e.*, ₹1055.71 per quintal in channel IV whereas it was ₹995.47 per quintal in channel III, ₹864.40 per quintal in channel II and ₹855.35 per quintal in channel I. The margin of the retailer was highest in channel I (₹614.00 per quintal) followed by channel III (₹540.00 per quintal) and channel II (₹510.00 per quintal). But the total marketing margin was highest in channel-II due to the presence of one more intermediary *i.e.*, wholesaler. These findings are in close conformity with Shah *et al.* (2010).

4.5.4.2.4. Marketing Efficiency in Different Channels

The results of the marketing efficiency of different marketing channels of Jammu district (Table 4.44) indicated that it was maximum (3.28) when farmer sold his produce directly to consumer *i.e.*, channel IV. When the fruit was sold through intermediaries *i.e.*, commission agent, wholesalers and retailers, marketing efficiency was less (0.37 in channel I, 0.39 in channel II and 1.16 in channel III). The results indicated that the marketing efficiency decreased with the increase in number of intermediaries. Further the marketing loss was nil or sometimes negligible in channel IV because orchardists used to sell his fruit crop immediately after harvesting and that too in nearby market, so loss due to perishability, handling and transportation was almost nil. Moreover, as far as marketing cost, marketing margin or net price received in channel IV, the producer got much more benefits as compared to other channels.

The results of the marketing efficiency of different marketing channels of Rajouri district (Table 4.45) indicated that it was maximum (4.46) in channel IV when farmer sold his produce directly to consumer. The marketing efficiency decreased with the increase in market intermediaries *i.e.*, more the marketing functionaries, less the marketing efficiency. It was 1.19, 0.91 and 0.79 in channel III, channel II and channel I, respectively. As far as marketing loss, marketing cost, marketing margin and net price received by the producer, similar pattern was observed in Jammu district.

The Kathua district of Jammu region also observed the similar trend with regard to marketing efficiency (Table 4.46). It was highest in channel IV (3.75) when farmer sold his produce directly to consumer.

The results of the marketing efficiency of different marketing channels of Samba district (Table 4.47) indicated that it was maximum (4.74) when farmer sold his produce directly to consumer *i.e.*, channel-IV. When the fruit was sold through intermediaries *i.e.*, commission agents and retailers, marketing efficiency decreases to 1.05 and 0.79 in channel-III and I, respectively, thereby indicating that the marketing efficiency decreased with the increase in market intermediaries *i.e.*, more the marketing functionaries, less the marketing efficiency. As far as marketing loss, marketing cost, marketing margin and net price received by the producer, similar pattern was followed as in case of Jammu district.

The results of the marketing efficiency of different marketing channels of Jammu region (overall) (Table 4.48) indicated that it was maximum (3.85) when farmer sold his produce directly to consumer. When the fruit was sold through intermediaries *i.e.*, commission agent, wholesalers and retailers, marketing efficiency was less (0.75 in channel I, 0.65 in channel II and 1.09 in channel III) thereby indicating that the marketing efficiency decreased with the increase in number of intermediaries. These results are in conformity with Ajani (2005) and Ladaniya *et al.* (2005). Further the marketing loss was nil or sometimes negligible in channel IV because orchardists used to sell his fruit crop immediately after harvesting and that too in nearby market, so loss due to perishability, handling and transportation was almost nil. Moreover, as far as marketing cost, marketing margin or net price received in channel IV, the producer got much more benefits as compared to other channels. So, if the government

policy is to make producer highest beneficiaries, channel IV is the most efficient marketing system.

4.5.5. Price Behaviour

4.5.5.1. Trends in Arrivals and Prices of Citrus in Narwal Market of Jammu District

The results pertaining to arrivals of orange in Narwal market (Table 4.49) indicated that the prices move contrary to arrivals. This showed that there was inverse relationship between arrivals and prices of orange in the sample market. The table also indicated the increase in arrivals and prices of orange over a period of time in the Narwal market. The peak market arrival period was from November with arrivals of 4093.00 qs to February with arrivals of 12339.40 qs. It was also observed that the prices were high in the month of May (₹3779.00 per q) as compared to other months and that was due to reduction in supply of oranges in relation to demand. These results are in close conformity with Vitonde *et al.* (1991).

The results pertaining to arrivals of kinnow in Narwal market (Table 4.49) indicated that the marketing season of kinnow started from the month of November every year and closed by the end of March. The arrivals were found to be highest (10116.00 qs) in the month of December and lowest in the month of March (2545.00 qs) while as the prices per quintal were lowest in the month of December (₹1308.60) and highest in the month of March (₹1829.20) thus depicted the inverse relationship between arrivals and prices of kinnow in the sample market during the period under reference. It can be inferred from above discussion that the increase of arrivals and prices of kinnow was lower as compared to orange.

The arrivals and prices of lemon (Table 4.49) in Narwal market indicated that the lemon remained available throughout the year and highest were recorded during the month of June (7293 qs) and lowest in December (1695.80 qs) whereas the prices were found to be lowest in December (₹1413.80 per q) and highest in March (₹2565.60 per q). The price of lemon changed in respect of its demand in the market and were lowest in December due to less demand. These findings are supported by Autkar *et al.* (1994).

4.5.5.2. Seasonal Fluctuations in Citrus Arrivals and Prices

It could be seen from the results of seasonal indices of arrivals and prices of orange, kinnow and lemon (Table 4.50) that a lean period in case of orange existed from May to June and September to November in Narwal market. The arrivals started picking up after November. During period from December to February arrival indices showed increasing trend (162.71 to 225.70) leading to glut in the market. Consequently the price index was lowest in the month of December (82.18). Thus, more supply was responsible for less price due to distress sale. It has been noticed that the prices started recovering from the month of January and reached the highest point during the month of May (137.00) and thus indicated the increase in the price index with the decrease in the arrival index. These results are in close conformity with Vitonde *et al.* (1991). The table also indicated that the situation in Narwal market for kinnow was slightly different as compared to orange as the kinnow in Narwal market came mainly

during the months of November to March. There were no arrivals of kinnow during the month of April to October. It was found that the prices (85.52) slump in the Narwal market due to higher volume of market arrivals (168.48) during the month of December. This shows that there was inverse relationship between arrivals and prices of kinnow in the sample market.

For lemon, the lean period existed from September to January and the arrivals started picking up after January. In the month of February and during period from May to July arrivals were the highest leading to glut in the market. Consequently, the prices were at their lowest ebb in the month of December (69.68). It was noticed that the prices started recovering from the month of January and reached the highest point during the month of March (126.45). It was also found that the prices slump in Narwal market due to higher volume of market arrivals during the month of June and July. These results are in close conformity with Anwarul Haq *et al.* (2006).

4.5.6. Constraints Faced by the Orchardists in Production and Marketing of Citrus

The various constraints both at production level and marketing level were worked out separately in the study area of Jammu, Rajouri, Kathua and Samba district of the Jammu region. Therefore, the constraints faced in general by the sample orchardists were lack of finance and credit facilities, inadequate irrigation facilities, non availability of good quality seedlings and farmyard manure, high cost of pesticides, non availability of labour during peak period, high cost of labour, occurrence of diseases, educated members go outside and did not take interest in farming and lack of latest technical knowledge and that of marketing of citrus *i.e.*, lack of processing units and marketing societies, lack of market information, un-organised marketing, low price paid to farmers, not getting remunerative price for the produce, less demand of fruits, problem of high perishability of fruits, problem of cheating in marketing by the middlemen, high and undue marketing margins and deductions in the market, costly packing material, high cost of transportation, non availability of market, high commission charges and packages not returned to the growers.

The results on major constraints faced (Table 4.51) by the orchardists of Jammu district, Rajouri district, Kathua district, Samba district and Jammu region as a whole in production were that of irrigation facilities (75.00 per cent, 96\5.83 per cent, 52.08 per cent, 39.58 per cent and 65.63 per cent respectively) and lack of finance and credit facilities (91.67 per cent, 89.58 per cent, 70.83 per cent, 79.17 per cent and 82.81 per cent, respectively) and hence farmers had difficulty because the good quality seedlings, FYM and pesticides were more expensive and unavailable which contributed to the additional cost of cultivation and are of opinion that the government fails to provide these inputs timely. The other constraint in fruit production were the occurrence of citrus diseases (41.67 per cent, 20.83 per cent, 12.50 per cent, 8.33 per cent and 20.83 per cent, respectively), lack of latest technical knowledge (58.33 per cent, 93.75 per cent, 39.58 per cent, 33.33 per cent and 56.25 per cent, respectively) and non-availability of labour (41.67 per cent, 25.00 per cent, 31.25 per cent, 52.08 per cent and 37.50 per cent, respectively) which severely limited actual production in spite of great inherent production. These findings are in close conformity with Phuse *et al.* (2008). Due to

either lack or high cost of transportation (31.25 per cent, 37.50 per cent, 60.42 per cent, 72.92 per cent and 50.52 per cent, respectively) and lack of market information (93.75 per cent, 85.42 per cent, 52.08 per cent, 20.83 per cent and 63.02 per cent, respectively), the fruits could not reach the right market at the right time, which compelled the fruit growers to sell the fruits in local market. Less demand of citrus fruit because of competition of other fruits resulted into low price received for the produce was another major marketing constraint. The problem of lack of processing units and co-operative societies expressed by 100 per cent respondents in Jammu and Kathua district and 95.83 per cent, 91.67 per cent, respectively in Rajouri and Samba district could have been the better option for the producers of the sample area to overcome the marketing problems upto certain extent but these units and societies were already facing shortage in the said area. These findings are supported by Sharan and Singh (2002).

Chapter 5
Future Strategies

The global citrus market is highly competitive and volatile. India's export markets and consumer demand for fresh citrus fruit as well as its juice consumption has increased dramatically. Also consumer understanding and knowledge of citrus is poor while as marketing has been under-resourced. Not only grower and packer numbers are declining as well as leadership and succession are critical issues. So, the vision of India should be for profitable citrus industry, which can delight its customers with high quality, great tasting products.

The same can be achieved if the following strategies are properly worked out:

5.1. Production Strategy

5.1.1. Product Improvement

Regional and varietal adjustment based on comparative advantage, regionalized and specialized production in compliance with regional comparative advantage, and biographic characters should be promoted. In the areas most suitable for growing citrus, early and late maturing varieties should be developed; wherever sweet orange can be grown, no loose-skin mandarin should be planted; year-round supply is to be realized in sweet orange producing areas; top quality with special features are required for medium maturing varieties. In respect of varietal structure, production of orange, especially the late varieties of top quality should be encouraged, with focus on juice-purpose varieties. Other varieties are appropriately encouraged, large-scaled planting for commercial purposes for lemon and kinnow in suitable areas should be advocated. Improvement of quality, balanced year-round availability of citrus fruit should be realized by varietal restructuring and introduction of high-quality varieties from abroad. And high quality at low cost should be achieved by reinforcing orchard management and increasing technical supports.

5.1.2. Research and Development

Research and development in the citrus fruits sector is needed to increase productivity as well as to improve resistance of the fruit to diseases and pests and for finding the best ways to fight those diseases and pests. Such type of research takes place in different areas for the improvement of the varieties and genetics, molecular biology and biotechnology, physiology of production and nutrition and quality of the fruits. In addition to this there should be search for new varieties that meet consumer demands of taste and appearance (such as easy peelers and fruits without seeds), as well as for earlier and long lasting crops in order to have round-year availability of the fruit and longer shelf life. So, R&D technologies for production should be promoted.

5.1.3. Harvesting Techniques

Productivity in the citrus growing fields can be improved by irrigation techniques and new fertilizing methods, as well as by mechanization processes for growing and harvesting the fruit.

5.1.4. Technical Know-How

Small and medium orchardists should be provided with the latest technical know-how, so as to improve level of yield for citrus. This will definitely push up their incomes. Therefore, extension of appropriate technology to the farmers for high-tech horticulture cultivation and precision farming is the need of the hour.

5.1.5. Replacement of Old Orchards

The procedure for the replacement of old orchards should be simplified and proper training should be provided for the purpose. This will help in replacement of old and sick orchards and production can be improved.

5.2. Marketing Strategy

5.2.1. Post-harvest Facility

Innovations in handling, packing, transport and storage techniques are useful to guarantee that the fruit and the juices arrive to the consumer at the highest levels of quality. Advances in refrigerating technologies are very important for the development of fruit transport and allow for citrus fruits to reach the consuming centers faster and with better quality. The new methods for transporting perishable technologies control better the old chain and allow for tracing the process through computerized systems.

5.2.2. Technology

Technological advances are among the major reasons behind the expansion in citrus fruits and citrus products production and trade. These technological improvements allow to better control the growing process, increase efficiencies at every stage of the marketing chain, from grower to retailer, and improve the quality and shelf life of the products. The evolution of transport and storage techniques has also been a catalyst for increasing the availability of citrus fruits all around the year and fruits from different origins can now be consumed almost everywhere.

5.2.3. Enhancing the Capability of Industry Units

Setting up of Industrial units for processing of citrus in Jammu and Kashmir is need of the hour. For this, building the confidence of the private investor in the horticulture development and fruit processing industry in this area is very important.

5.2.4. Value Addition to Citrus By-products

The government should assist for setting up post harvest facilities such as pack house, cold storages, controlled atmosphere storages etc, processing units for value addition and marketing infrastructure.

5.2.5. Increase Consumer Demand for Citrus Juice

Demand depends on factors such as income levels, population growth, availability and relative prices of substitute fruits and the changing consumer preferences for fresh produce, including health, quality, convenience or taste characteristics. Promotion campaigns may play an important role in order to increase demand for citrus fruits and juices.

Bibliography

Acharya, S. S. and Aggarwal, N. L. 2001. *Agricultural Marketing in India*. Third edition, Oxford and IBH Publishing Company, New Delhi.

Ahmad, B. and Mustafa, K. 2006. Forecasting kinnow production in Pakistan: An econometric analysis. *International Journal of Agriculture and Biology*, **8**(4): 455-458.

Ajani, O. I. 2005. Economic analysis of the marketing of fruit in Lagos State of Nigeria (A case study of Oyingbo, Oshodi and Ikotun markets). *Nigerian Journal of Horticultural Science*, **10**: 38-46.

Anonymous, 2008. State-wise area, production and productivity of fruits. APEDA. Available at www.apeda.com.

Anonymous, 2009. Economic Survey, Directorate of Economics and Statistics, Ministry of Agriculture, Govt. of India.

Anonymous, 2009. Economic Survey, Directorate of Economics and Statistics, Planning and Development Department, Government of Jammu and Kashmir.

Anonymous, 2011a. Economic Survey, Directorate of Economics and Statistics, Ministry of Agriculture, Govt. of India.

Anonymous, 2011b. Achievements of Agriculture production Department.

Anonymous, 2010. Available at www.marketpublisher.com.

Anwarul Haq, A. S. M., Matin, M. A. and Habibul Haque, A. K. M. 2006. Kagzi lemon marketing system in selected areas of Bangladesh. *Economic Affairs*, **51**(1): 14-23.

Auhurkar, B. W. and Deole, C. D. 1985. Producers' share in consumers' rupee: A case study of fruit marketing in Marathwada. *Indian Journal of Agricultural Economics*, **40**: 403.

Autkar, V. N., Vyawahare, C. A. and Koranne, V. M. 1994. Scope for development of kagzi lime processing industry in Vidarbha. *The Bihar Journal of Agricultural Marketing*, **2**(1): 73-77.

Bagde, N. T., V. N. Autkar and C. A. Vyawahare. 1996. Dynamics of marketing of selected fruits in Nagpur district. *The Bihar Journal of Agricultural Marketing*, **4**(1): 71-77.

Bakhsh, Khuda, Hassan, Ishtiyaq and Akhter Muhammad Shafiq. 2006. Profitability and Cost in Growing Mango Orchards. *Journal of Agriculture and Social Sciences*, **2**(1): 46-50.

Banerjee, G. D. 2009. Poised for a golden revolution. *Times Agriculture Journal*.

Bhole, B. D. 2004. Problems in marketing of oranges in Vidarbha. Research project. Available at www.google.com.

Braddock, R. J. 1999. Handbook of citrus by-products and processing technology.

Chavan, V. S. 2004. *Marketing of pomegranate in Sangli district of Maharashtra*. M.Sc (Agri.) thesis. MPKV, Rahuri, India.

Chinnapa, B. and Rammana, R. 1997. An economic analysis of guava production. *Agricultural Banker*, **21**(3): 29-33.

Choubey, Manish and Atteri, B. R. 2000. Economic evaluation of litchi production in Bihar. *The Bihar Journal of Agricultural Marketing*, **8**(2): 123-131.

Economos, C. and Clay, W. D. (1999). Nutritional and health benefits of citrus fruits. *FAO, Food, Nutrition and Agriculture*, **24**.

Gangwar, L. S., Ilyas, S. M., Singh, D. and Kumar, S. 2005. An economic evaluation of kinnow mandarin cultivation in Punjab. *Agricultural Economics Research Review*, **18**(1): 71-80.

Gangwar, L. S., Singh, Shyam. 1998. Economic evaluation of Nagpur mandarin cultivation in Vidarbha region of Maharashtra. *Indian Journal of Agricultural Economics*, **53**(4): 648-653.

Ghafoor, U., Muhammad, S., Chaudhary, K. M., Mahmood, A. R., Ashraf, I. 2010. Harvesting and marketing problems faced by citrus (kinnow) growers of tehsil Toba Tek Singh. *Journal of Agricultural Research*, **48**(2): 253-257.

Gill, K. S. 1980. Marketing problem of citrus grown in the Punjab state. Research report, Department of Economics and Sociology. PAU, Ludhiana. pp 1-42.

Gorinstein, S., Martin-Belloso, O., Park, Y., Haruenkit, R., Lojek, A., Milan, I., Caspi, A., Libman, I. and Trakhtenberg, S. 2001. Comparison of some biochemical characteristics of different citrus fruits. *Food Chemistry*, **74**(3): 309-315.

Gummagolmath, K. C., Hiremath, G. K. and Kulkarni, Basavaray. 2002. Mango cultivation in Dharwad district: An analysis of cost and resource productivity. *The Bihar Journal of Agricultural Marketing*, **10**: 148-152.

Gupta, G. S. and George, P. S. 1974. Profitability of Nagpur santra (oranges) cultivation. *Indian Journal of Agricultural Economics*, **29**(1): 134-142.

Hanumantharaya, M. R., Kerutagi, M. G., Patil, B. L., Kanamadi, V. C. and Bankar, B. 2009. Comparative economic analysis of tissue culture banana and sucker propagated banana production in Karnataka. *Karnataka Journal of Agricultural Sciences*, **22**(4): 810-815.

Heathfield, D. F. and Wibe, S. 1987. An Introduction to Cost and Production Functions. Macmillan.

Holton, R. 1953. Marketing Structure and Economic Development. Quarterly Journal of Economics. **67**: 344-361.

Hugar, L. B., Murthy, P. S. S., Umesh, K. B. and Reddy, B. S. 1991. Economic feasibility of Guava cultivation under scientific management – An Enperical Evidence. *Agricultural Situation in India*, **46**(4): 211-214.

Iqbal, Mudasir. 2009. *Investment appraisal of mango and ber fruit production in Jammu district of J and K state.* M.Sc. thesis. Sher-e-Kashmir University of Agricultural Sciences and Technology of Jammu, Jammu, India.

Ismail, M. and Zhang, J. 2004. Postharvest citrus diseases and their control. *Outlook Pest Management*, **1**(10): 29-35.

Kachroo, J. 2004. Growth, profitability and projection of major fruit crops in J and K state, India. *The Bangladesh Journal of Agricultural Economics*, **27**(1): 81-94.

Kadam, A. L., Chaudhari, M. R., Nagpure, P. M. and Lanjewar, A. D. 2000. Constraints in marketing management of oranges faced by farmers. *Journal of Soils and Crops*, **10**(2): 286-288.

Khushk, A. M., Memon, A. and Lashari, M. I. 2009. Factors affecting guava production in Pakistan. *Journal of Agricultural Research*, **47**(2): 201-210.

Kimball, D. A. 1999. *Citrus Processing: A Complete Guide* (2nd Ed.). Springer. pp. 9.

Koujalgi, C. B., Poddar, R. S. and Kiresur, V. R. 1999. Profitability of production of marketing of pomegranate orchards in Bijapur district, Karnataka. *Indian Journal of Agricultural Economics*, **58**(4): 811.

Kumar, S., Karol, A., Singh, R. and Vaidya, C. S. 2007. Cost and return from apple cultivation: A study in Himachal Pradesh. *Agricultural Situation in India*, **64**(1): 307-313.

Ladaniya, M. S., Vinod Wanjari and Mahalle, B. C. 2003. Price spread of pomegranate. *Indian Journal of Agricultural Economics*, **58:** 800-811.

Ladaniya, M. S. and Wanjari, V. 2003. Marketing pattern of mosambi (sweet orange) in selected districts of Maharashtra. *Indian Journal of Agricultural Marketing*, **17**(1): 52-62.

Ladaniya, M. S., Wanjari. V. and Mahalle, B. C. 2005. Marketing of grapes and raisins and post-harvest losses of fresh grapes in Maharashtra. *Indian Journal of Agricultural Research*, **39**(3): 167-176.

Langde, V. V., Pawar, B. R., Deshmukh, D. S. and Yeware, P. P. 2010. Resource productivity and resource use efficiency in flood irrigated banana production. *International Journal of Commerce and Business Management*, **3**(1): 114-116.

Lepeha, Y., Ali Md. H., Maity, A., Mukherjee, A. K. and Chattopadhayay, T. K. 1993. Economics of Marketing of mandarin orange in Darjeeling district of West Bengal. *Economic affairs*, **38**(4): 232-241.

Mali, B. K., Bhosale, S. S., Shendage, P. N., Shindh, C. B. and Kale, P. V. 2004. Economics of production and marketing of Banana in Jalagaon district of western Maharashtra. *Agricultural Situation in India*, **60**(11): 733-741.

Mittal, Surabhi. 2007. Can horticulture be a success story for India? Indian Council for Research on International Economic Relations. Working Paper No. 197. 1-79.

Mondal, A. H. 1986. Marketing of pineapple in Meghalaya state. M.Sc. thesis. Department of Economics and Sociology. PAU, Ludhiana.

Mruthuyunjaya 2001. Marketing of Fruits and Vegetables: Some Issues and Suggestions in (Ed.) Ajit Singh. *Problems of Small Marginal farmers in Marketing fruits and Vegetable*. Books India International, New Delhi: 1-10.

Naikwadi, D. J., Sabale, R. N. and Khaire, V. A. 2004. Economics of production and marketing of fig in Pune district. AGRESCO. Report, Deptt. of Agricultural Economics. MPKV, Rahuri. 199-213.

Nandal, R. S. and Punia, Deep. 2004. Economics of major fruit crops: A study of western zone of Haryana. *Indian Economic Panorama*, **14**(3): 53-56.

Ozkan, B. Akcaoz, H. V., Karadeniz, C. F. 2002. Costs and returns of citrus production in Antalya province. *Ziraat Fakultesi Dergisi, Akdeniz Universitesi*, **15**(1): 1-7.

Patel, S. P. and Pawar, J. R. 1980. A study of marketing of fruits in Mahatama Phule market, Bombay. *Horticultural Economic Newsletter*, pp 231-247.

Phuse, A. R., Atkare, P., Vitonde, A. K. and Wankhade, R. S. 2008. Constraints and suggestions in Nagpur mandarin orange production. *Journal of Soils and Crops*, **18**(2): 417-421.

Radha, Y., Krishnaiah, J. and Prasad, Y. E. (2000. Economic appraisal of citrus project: A case study. *Journal of Research. ANGRAU*, **28**(3): 12-18.

Radha, Y. and Prasad, D. S. and Reddy, S. J. 2006. Economic analysis and production and marketing of grape in Andhra Pradesh. *Indian Journal of Agricultural Research*, **40**(1): 18-24.

Reddy, Y. V. R. 1987. Case studies in Horticulture Project Evaluation. *Indian Journal of Agricultural Production Economics*, **42**(3): 623-625.

Satihal, D. G. 1993. *Economics of production and marketing of ber in Bijapur district, Karnataka*. M.Sc. Thesis. The University of Agricultural Sciences, Dharwad.

Shah, N., Khan, M., Khan, N., Muhammad Idrees, Haq, I. 2010. Profit margins in citrus fruit business in Haripur district of NWFP, Pakistan. *Sarhad Journal of Agriculture*, **26**(1): 135-140.

Sharan, S. P. and Singh, V. K. 2002. Marketing of kinnow in Rajasthan. *Agricultural Marketing*. **45:** 2-4.

Sharif, M., Akmal, N. and Taj, S. 2009. Financial viability for investing in citrus cultivation in Punjab, Pakistan. *Journal of Agricultural Research (Lahore)*, **47**(1): 79-89.

Sharma, D. K., Singh, V. K. Khatkar, R. K. and Sharma, S. 2006. An economic analysis of mango cultivation in Yamunanagar district of Haryana. *Haryana Agricultural University Journal of Research*, **36**(2): 105-111.

Sidhu, M. S. 1993. Price spread of kinnow in Punjab. *Indian Journal of Agricultural Marketing*, **7**: 105-113.

Sikka, B. K., Singh, Ranveer and Kumar, Rakesh. 1992. Profitability of apple cultivation in Himachal Pradesh. *Agricultural Situation in India*, **47**(8): 637-640.

Singh, N. 1987. Economics of production and marketing of horticultural crops. *Indian Journal of Agricultural Production Economics*, **42**(3): 620-621.

Singh, V. and Khatkar, R. V. (1994). An economic analysis of grape cultivation in Hissar district of Haryana state. *Haryana Agricultural University Journal of Research*, **24**(4): 176-183.

Singh, S. K. and Sayeed, Mohammad. 2008. Economic analysis of cultivation of aonla in Sandawa Chandrika block of Pratapogarh district of Uttar Pradesh. *Indian Journal of Agricultural Production Economics*, **63**(3): 369.

Singh, R. and Sikka, B. K. 1992. Production and marketing system of apple in tribal area of Himachal Pradesh. *Agricultural Situation in India*, **47**(8): 595-597.

Spiegel, Roy and Goldschmidt, E. E. 1996. *Biology of Citrus*. Cambridge University Press. pp. 4.

Subrahmanyam, K. V. 1986. Profitable lime cultivation in Andhra Pradesh. *Indian Horticulture*, **31**(2): 25-28.

Subrahmanyam, K. V. 1987. Economics of Investment in mango cultivation in Karnataka. *Mysore Journal of Agricultural Science*, **21**: 196-200.

Sudha, M. and Reddy, Y. V. R. 1988. Economics of sweet orange cultivation. *Indian Horticulture*, **33**(3): 24-25.

Supe, M. S., Kale, M. U. and Katkhede, S. S. 2009. Cost economics of kagzi lime production with different irrigation systems - a case study. *Green Farming*, **2**(6): 386-387.

Suresh, A. and Reddy, T. R. K. 2004. An economic analysis of Banana cultivation in Peechi command area of Thrissur district of Kerala state. *Agricultural Situation in India*, **61**(9): 629-631.

Tewari, S. C. 1994. Income and investment of vegetable and fruits growing farms – a comparative study. *Himachal Journal of Agricultural Research*, **10**: 40-44.

Thakur, D. S., Saibabu, M. V. S. and Sharma, K. D. 1986. Economics of kinnow cultivation in Himachal Pradesh. *Himachal Journal of Agricultural Research*, **12**(1): 47-51.

Thilagavathi, M., Siddeswaran, K. and Alagumani, T. 2002. Economic viability of dryland horticulture in rainfed vertisols – A study in Southern Tamil Nadu. *Agricultural Economic Research Review. (Conference Issue)*, 58-62.

Tomer, B. S., Singh, S. P. and Khadhar, R. K. 1997. Marketing of grapes and citrus fruits in Haryana. *Indian Journal of Agricultural Economics,* **52:** 642.

Utomakili, J. B. and Molua, E. 1998. Analysis of resource use efficiency in banana Musa Sp. (AAA group) production in the south-east province of Cameroon: A case study. *International Journal of Tropical Agriculture,* **16**(1-4): 113-118.

Vitonde, A. K. and Bhargawa, P. N. 1991. Marketing of mandarin orange in Nagpur district. *Indian Journal of Agricultural Marketing,* **5**(2): 212-215.

Wagale, S. A., Talathi, J. M., Naik, V. G. and Malave, D. B. 2007. Resource use efficiency in Alphonso mango production in Sindhudurg district (M.S.). *International Journal of Agricultural Sciences,* **3**(1): 28-34.

Wani, G. M. 2008. Past, Present and Future of Horticultural Development in J and K, India. Available on www.buzzle.com.

Wani, M. H., Singh, R. I., Bhat, A. R. and Mir, N. A. 1993. Resource use efficiency and factor productivity in apple. *Agricultural Economic Research Review,* **6**(1): 26-31.

Webber, H. John. 1967. *History and Development of the Citrus Industry*. University of California, Division of Agricultural Sciences. http://lib.ucr.edu/agnic/webber.

Yeware, P. P., Pawar, B. R., Landge, V. V. and Deshmukh, D. S. 2010. Costs, returns and profitability of *Mrugbahar* and *Ambebahar* sweet orange production. *International Journal of Commerce and Business Management,* **3**(1): 79-81.

Yotopoulos, P. A. and L. J. Lau. 1979. Resource Use in Agriculture: Applications of the Profit Function to Selected Countries. *Food Research Institute Studies.* **17**(1):1 – 115.

Zaman, Q. and Schumann, A. 2005. Performance of an Ultrasonic Tree Volume Measurement System in Commercial Citrus Groves. *Precision Agriculture,* **6**(5): 467 – 480.

Index